CONTENTS

[4]

[5]

PREFACE

The *Hobart Papers* are intended to contribute a stream of authoritative, independent and lucid analyses to the understanding and application of economics to private and government activity. The characteristic theme has been the optimum use of scarce resources and the extent to which it can best be achieved in markets within an appropriate framework of law and institutions or, where markets cannot work, in other ways. Since in the real world the alternative to the market is the state, and both are imperfect, the choice between them effectively turns on judgement of the comparative consequences of 'market failure' and 'government failure'.

In the autumn of 1979 a major Public Inquiry, lasting for some six months, began into a proposal by the National Coal Board to mine coal in North East Leicestershire, principally in the Vale of Belvoir. It was recognised by both advocates and opponents that the first priority of the Inquiry was to establish whether or not there was a 'need' for Belvoir coal. (The question-begging term 'need' is widely used by non-economists in a vague sense but requires very close examination to see whether it has any economic content.) In the words of the NCB counsel's closing statement:

> 'We would unreservedly agree with Sir Frank Layfield (counsel for Leicestershire) when he said . . . "Economic need is the first factor to be considered".'

The Inquiry, therefore, had to discuss and criticise the Board's plans for expanding coal output in Britain, as well as the specific proposal to sink three pits in and around the Vale of Belvoir.

Professor Colin Robinson appeared as a witness for Leicestershire County Council (which opposed the NCB's proposal) and argued that the NCB had failed to produce convincing evidence for its expansion plans. In general, he suggested, they were predominantly 'supply determined', aiming at reaching quantities of output not specifically related to likely consumer demand given probable trends in the prices of coal and other fuels.

In Hobart Paper 89 Professor Robinson and Eileen Marshall

[7]

analyse the prospects for the British coal industry, based originally on research carried out for the Belvoir Inquiry but brought up to date in the light of events since the Inquiry ended in the Spring of 1980. The *Paper* discusses the market for coal in Britain, where it has been restricted by the government creation of virtual monopoly, and in the world market, which is more competitive. The analysis includes the most recent developments in February-March 1981 in which the use of the strike-threat by the miners' union induced the Government to yield to its demands. The *Hobart Paper* thus analyses competition and monopoly in the market both for coal and for the labour that produces it, the consequences of nationalisation in Britain in a competitive world market, and the lessons that may have to be learnt if a politically-created monopoly is not to resist the adaptation of industry to changing conditions of supply and demand.

Professor Robinson, a leading authority on the economics of the energy industry, and Eileen Marshall have produced a model *Hobart Paper* in applying the main elements of economic theory to an industry in which political bargaining has tended to obscure the opportunity costs of monopoly to other industries, to domestic consumers and to the economy as a whole. They trace the origins of the recent events for several decades in which the market for British coal has declined with technological innovation and rising living standards. In so doing they illustrate the power of market analysis to illuminate the changing structure and fortunes of the coal industry and of its employees at the coal face and above ground. Above all they demonstrate the damaging consequences of the nationalisation that, unlike the competitive production of coal in large producing countries overseas, has reduced the impact on British coal of changing market conditions that would have led it to make its adaptations more gradually and so avoided the discomfort and dislocations that now have to be made when the adaptations to changing conditions can no longer be postponed. Although the industry has been reduced in size as newer forms of fuel have emerged, the rate of adaptation has been determined not primarily by the decisions of consumers choosing between coal and newer fuels, but by the capacity of the producers to resist reforms and to slow them down to suit themselves.

The authors present an alarming portrayal of the ability of the National Coal Board, the National Union of Mineworkers and even the Department of Energy, which is supposed to safe-

[8]

guard the consumer interest above all, to use unrealistically optimistic forecasts of the demand for coal, based on low-quality economic argument, in order to induce government to allocate scarce resources to coal production that might have been used more effectively elsewhere. And these romantic forecasts have also sadly raised false hopes among the miners, whom they were designed to protect. But that is the consequence of creating a monopoly in which the producer is judge and jury of his cause. The authors present their own estimate that the consumption of British coal will have dwindled by a further 11-40 per cent by the year 2000.

Professor Robinson and Eileen Marshall emphasise the distinction between the markets in British coal and for world coal. They argue that the higher costs of nationalised British coal contrast with the lower costs of privately produced coal in the USA, South Africa and Australia, and that this contrast largely accounts for the relatively poor outlook for British coal and the relatively bright outlook for overseas coal. Their conclusion is that there is little or no case for further protection of nationalised coal from the competition of cheaper and/or better coal from overseas, which would be to the advantage of the British steel, electricity and other industries, and of the householder.

The *Hobart Paper* raises disturbing fundamental issues in the conduct of British public policy. The claim made for nationalisation 35 years ago was that it would ensure that industry would be run for the good of the community as a whole. Little thought was given to the ability of nationalised industries to adapt themselves to adverse market conditions at home and overseas. The coal industry now illustrates the high price that is being paid for enabling nationalised industry to exploit the community by threatening government by local or general strike. Here, as elsewhere, there will have to be more attention paid to the economics of 'public choice', which studies decisions made without the assistance of markets, and to the reforms in the British constitution to deny transient government the power to yield to vested interests at the expense of the general interest. It also indicates the neglected task of how to take out of governmental control and political preferment industries that do not have to be owned, financed or run by government, or its agencies, because they are not public goods and there are no large-scale economies to justify centralised control. The question is whether the adaptations would have been made with less traumatic upheaval if the industry had, as elsewhere, been in private ownership spending its own rather than taxpayers' money.

Professor Robinson and Eileen Marshall have written an authoritative, scholarly, clearly argued, and spirited analysis that will evoke respected attention from a wide range of readers from teachers and students of economics to people in public life responsible for making or influencing policy. The Institute's constitution requires it to preclude its Trustees, Advisers and Directors from necessarily accepting the analysis or conclusions of its published work, but it presents this *Hobart Paper* as a timely, persuasively argued and disturbing dissection of an industry that has caused British economy and society grievous anxiety and suffering for over half a century since the General Strike of 1926.

April 1981   ARTHUR SELDON

THE AUTHORS

Colin Robinson was born in Stretford, Lancashire and educated at Stretford Grammar School and Manchester University, from which he graduated in 1957 with a First in Economics. For the next eleven years he worked as a business economist, first with Thomas Hedley and then with Esso Petroleum Company and Esso Europe (where he was economic adviser during the North Sea gas negotiations). In November 1968 he was appointed to the chair of economics in the University of Surrey, where a substantial research group in energy economics has been built up.

Professor Robinson has written a book, *Business Forecasting: An Economic Approach* (Nelson, 1971), and several papers on business forecasting. Numerous works on the economics of the fuel 'industries and on fuel policy include *A Policy for Fuel?* (Occasional Paper 31, IEA, 1969), *Competition for Fuel* (Supplement to Occasional Paper 31, 1971), *The Energy 'Crisis' and British Coal* (Hobart Paper 59, IEA, 1974), *North Sea Oil in the Future*, Macmillan, 1978 (with Jon R. Morgan), and *The European Energy Market in 1985*, Staniland Hall, 1978 (with George F. Ray). He is joint editor of *Energy Economics,* a member of the editorial board of *Energy Policy*, and a member of the Electricity Supply Researxh Council.

Eileen Marshall took a First Class Honours degree at the University of Surrey where she is now a Lecturer in the Department of Economics teaching applied economics and environmental economics. Her research is principally in the economics of energy conservation. Before her present appointment she spent some years working in the City as a stockbroker.

PROLOGUE

The main concern of this *Hobart Paper* is to analyse the future of the British coal industry to the end of the century. Contrary to popular (though unsupported) opinion, our view is that the prospects are not good. We recognise the enormous difficulty of looking into the future of energy, and we make no pretence of forecasting developments with any accuracy. We try instead to set out some probable broad trends and ranges of uncertainty.

The analysis (assuming no increase in government subsidies) leads us to believe that, at best, a slight fall in British coal output is likely by the end of the century. Quite probably a more substantial decline will occur. It seems to us that the future of coal worldwide lies in a number of low-cost producing countries (for example, the United States, Australia and South Africa) and that, for the benefit of British consumers, there should be freedom to import coal into this country. We argue that import restrictions would increase the monopoly power of the British coal industry and that, as a consequence, the price of coal in Britain would tend to follow the rising price of oil. Moreover, increased dependence on indigenous coal might well lead to more interruptions in fuel supplies.

The power of monopoly?

Since this *Paper* was prepared there has evidently been a change of government policy following a remarkable practical demonstration of the monopoly power we analyse. In mid-February 1981, the National Coal Board (NCB) and the National Union of Mineworkers (NUM) forced the Government into a change of policy which, according to reports, will mean the continued operation of presently uneconomic pits, more subsidisation of coal, and fewer coal imports.

In January and early February 1981 there were reports that pit closures were to be accelerated. Then, on 10 February, the NCB let it be known to the mining unions that it would probably want to close between 20 and 50 pits in the next five years, claiming that its problems resulted from the recession and tight financial constraints imposed by the Government.

[13]

Between 20,000 and 30,000 jobs would be affected by the closures, though the number of redundancies would be fairly small because of natural wastage, transfers within the industry, and voluntary premature retirements.[1] Within a week the miners reacted in an outbreak of unofficial strikes in Wales, Scotland and Kent, and a nationwide coal strike against the closure programme seemed highly probable, with possible sympathetic action by other unions. Since the Government was apparently unwilling to face such a strike, it agreed in principle on 18 February to give more money to the industry. Thereupon the NCB withdrew its closure proposals, the NUM called off its threatened strike ballot, and unofficial strikes quickly ceased.

Unrealistic forecasts to bolster state aid

During the 1960s and early 1970s the NCB had contended that it required more government aid in the short run to survive into the better long-term future it claimed for British coal. More recently, it has based similar claims on forecasts and targets which the analysis in our *Paper* suggests are unrealistic; and it has had some qualified approval from the Government for its expansion plans.

If our analysis is anywhere near correct, both the NCB and recent governments have seriously and consistently misled employees in the coal industry, encouraging expectations about their prospects which were unlikely to be fulfilled in the foreseeable future. Naturally enough, therefore, when the Board suddenly began in late 1980 and early 1981 to lay less emphasis on its supposedly bright long-term future and instead announced large numbers of planned pit closures, disappointed expectations led to an extreme reaction from the miners. On 18 February 1981 a show of force apparently succeeded where argument had previously failed.

In mid-March 1981 the form of the new policy was still not entirely clear. The Budget of 10 March provided for grants to industry of up to 25 per cent of the capital cost of boiler conversions to coal-firing; the Chairman of the NCB, Sir Derek Ezra, evidently suggested after the Budget that the cost of avoiding pit closures and subsidising British coal to reduce

---

[1] E.g. 'Miners prepare for fight to prevent more pit closures', *Financial Times*, 11 February 1981.

imports would be between £100 and £200 million in 1981-82. The President of the NUM, Mr Joe Gormley, is reported to have asked for further aid in the form of lower interest payments by the NCB, financial assistance for stocking and exporting coal, and government support for a planned coal liquefaction plant in Wales.[1]

Thus the coal industry still appears to be pressing home the advantage it won in February. In Mr Gormley's words: 'The commitment we must have is that the country will sell every ounce of coal we produce.'[2] Whether fuel consumers will respond by taking more British coal is uncertain. The Central Electricity Generating Board (CEGB) already has what is effectively an import-limiting agreement with the NCB. Neither the CEGB nor the British Steel Corporation (BSC) is likely to welcome an arrangement which reduces the limited competitive pressure on the Coal Board from small quantities of foreign coal. As our *Paper* explains, relatively low-price imports have been enlarging their share of the British market, and it will not be easy to reduce them except by explicit import-licensing. Consumers in general will doubtless be concerned at the display of power which has forced the change in government policy and will conclude that what has happened could occur again. They may well have reservations about switching to a fuel in which there is a powerful domestic monopoly, with all that may imply in terms of high prices and vulnerability of supplies.

Justification for short-term gain from new policy?
Thus the short-term gain from the new policy, even to the coal industry, may be small, and there may be little or no longer-term benefit to the industry. What justification might there be on wider grounds for the policy change? Even though it results from unwillingness to upset a powerful pressure-group, it could conceivably be supported by economic or other arguments.

We can dismiss right away one commonly-expressed argument for increased protection—that we should match the higher direct subsidies which some other European countries (such as Germany, France and Belgium) give to their coal

---

[1] 'Coal industry aid will top £100m', *Financial Times*, 12 March 1981.

[2] Quoted in 'NCB and miners agree on united front', *Financial Times*, 24 February 1981.

[15]

industries. Apart from potentially damaging consequences for international trade of beggar-my-neighbour protection for coal, even on nationalistic grounds it makes no sense merely to ape what others choose to do. It is necessary to think of policies which are justified by our own circumstances. Furthermore, there is a good deal of indirect coal protection in Britain, such as the tax on fuel oil (£8 per tonne) which is higher than in other large West European countries, and government pressure on the CEGB and BSC to use British coal.

A possible economic rationalisation for the change of policy relates to the social costs of pit closures. In economic terms, it can be argued that 'shadow wages' in mining are well below money wages since some miners would find difficulty obtaining work elsewhere. This social-cost argument invites several comments. *First,* it would be naïve to believe that the Government has explicitly considered the social-cost arguments and is carefully calculating its aid accordingly. Indeed, it is inherently implausible that the outcome of a very brief bilateral monopoly bargaining session, in which one side was under strike threat, would in any way approach what is socially desirable. It is much more likely that the combined weight of the NUM and the NCB would bias the outcome very much in their favour.

Second, it is not clear that the pit closures would have left most of the affected miners unemployed. As already mentioned, the NCB appears to have stated that few redundancies would have resulted. Under the new arrangement, the NCB will presumably operate less efficiently than with its preferred pit closures because it will not be able to re-deploy its labour force as it wished. In such circumstances, it is by no means certain that mining employment will be expanded by the change of policy, except in the short term.

Third, if the aid really was agreed to on social-cost grounds, it would seem curious to single out coal mining for special treatment when there are other industries in temporary or permanent decline to whose employees exactly the same principles could be applied. Indeed, given the recent big decline in manufacturing employment in Britain, compared with the near-stability of the labour force in mining, the miners' case for special aid does not look at all strong. There are already many regions and communities where the social costs of local factory closures are high. If the shadow-wage argument was pursued

[16]

to justify similar action in such areas to that now being taken in mining, there would be a massive programme of output support for British industries and wholesale explicit or implicit import restrictions (which would be very damaging in the long run because of its effects of increasing costs and slowing-down structural change). The Government has not suggested such a programme; it can only be assumed that the coal industry has been selected for special treatment because of the industrial might it has so amply demonstrated. For industrial and domestic consumers the outcome looks much less happy since they may well be faced with higher coal and electricity bills as a consequence of increased monopolisation of the British fuel market. There will also be smaller funds available to help lower-income groups who will be particularly hard-hit by increased fuel charges.

There is a case for tempering the effects of declining employment in mining as in other industries, but it seems more logical to us to help *people* who become unemployed rather than to support *output*[1]—especially since this policy would tend to strengthen the coal industry's monopoly power. The increase in redundancy payments for coal industry employees recently announced by the Government[2] seems a sensible way of dealing with the real income losses resulting from a fall in mining employment. No doubt more could also be done to remedy imperfections in the labour market by improving mobility and re-location allowances and re-training programmes. Such aid for people has in-built time limits (for instance, as they are re-employed or retire) and is also a much more explicit charge (which can be re-considered from time to time) on the community than the concealed cost of sustaining uneconomic output indefinitely.

An open-ended government commitment?

As a whole, the Government's new policy seems perilously close to an open-ended commitment to support production at any cost. What, for instance, will happen if consumers do not turn to coal despite subsidies? Will the subsidies be increased on the ground that pits cannot be closed unless there are insurmountable geological problems? All reasonable people will condemn

[1] Colin Robinson, *A Policy for Fuel*, Occasional Paper 31, IEA, 1969.
[2] 'Coal industry aid will top £100m', *Financial Times, op. cit.*

the wastefulness and human misery of unemployment but that is a different matter from attempting to support production at a figure and in a way which almost certainly cannot be sustained economically in the long run.

In summary, the Government's change of course seems to us a prime example of surrender to a powerful producer-group with little regard to the interests of society as a whole. No matter how much sympathy there may be for the coal industry's employees, particularly since they appear to have been misled about their likely future, it can only be concluded that the Government is placing the British coal consumer in the hands of an indigenous monopoly. That is not the way to keep down our energy prices in the long run, nor to promote the security of our fuel supplies.

I. RECENT DEVELOPMENTS

The NCB's plans in perspective

To appreciate the significance of the NCB's proposals for expansion, it is essential to view them in the context of the recent history of British coal. We begin, therefore, by explaining the scale of the Board's plans for the future. Then we give a brief account of historical trends in the industry, concentrating on the post-war period, followed by a review of government attitudes towards coal.

One difficulty at the outset is an element of uncertainty about the present status of the NCB's published plans. The most recent document which defined those plans was *Coal for the Future*,[1] published as long ago as February 1977 by a 'Tripartite Group' consisting of the NCB, the coal trade unions and the Ministers concerned. It set out a programme, generally known as 'Plan 2000', which aimed to raise British coal output from 124 million tonnes in 1976 to 170 million tonnes in the year 2000 (a 37 per cent increase). Of the projected total, deep mines were to produce 150 million tonnes and open-cast operations the other 20 millions.

Since there has been no published revision to Plan 2000, we take it as the best broad indication available of the Coal Board's plans for the end of the century. However, the Board's witnesses at the Belvoir Inquiry were somewhat evasive when questioned about whether Plan 2000 still stands in its entirety. One change which emerged during the Inquiry was that end-century open-cast production is now expected to be 15 million tonnes rather than 20 millions. Possibly there has been some downward revision also of the 150 million tonne target for deep-mined coal in the year 2000, but, since no change has been made public, we assume in this *Hobart Paper* that the 150 million tonne figure still stands, at least as a general aspiration.

---

[1] Department of Energy, *Coal for the Future*, 1977.

A summary of the principal features of the British coal market in recent years will illustrate the magnitude of the reversal in coal's fortunes implied by the NCB's plans for the future.[1]

There was a very pronounced downward trend in British coal production from the early years of this century up to the mid-1970s (Table I). In 1913, the industry achieved its peak output of 287 million tonnes, of which about one-third was exported. Subsequently, however, despite occasional revivals (as in the late 1940s and early 1950s), home consumption fell drastically and exports became very small. During the mid-1970s, output and home consumption were in the range of 120 to 130 million tonnes a year (40 to 45 per cent of 1913 output) and net exports fluctuated around zero.

From 1957 until the mid-1970s British coal was in particularly rapid decline (Table I). In 1977 deep-mined production was only about half what it had been in 1957 and mining employment was approximately one-third of its 1957 figure. Total coal production declined proportionately less than deep-mined output since open-cast output was about the same in 1977 as in 1957. The drop in inland consumption was of approximately the same size as the fall in total production because imports and exports of coal are small (though coal exported and supplied to foreign ships' bunkers in 1977 was only a quarter as much as in 1957). The relatively much larger reduction in employment than in output (Table I) is associated with the big productivity increase of the 1960s. From 1957 to 1971 output per man at NCB mines rose from 301 to 478 tonnes (nearly 60 per cent), but output per man in 1977 was down to 442 tonnes. Many less productive pits have been closed since 1957 so that the number of NCB mines, 822 at the end of 1957, was 289 in 1971 and 231 in 1977.

Between 1977 and 1980 there were signs of more stability in the industry. Production rose about 5 per cent, productivity improved (though output per man-year was still lower than in 1971), employment fell only slightly, and consumption was essentially unchanged (Table I).

Within the 1977-80 period, however, there were substantial

[1] The post-war history of coal and government policy towards the industry are discussed in Colin Robinson, *A Policy for Fuel?*, Occasional Paper 31, IEA, 1969; *Competition for Fuel*, Supplement to OP 31, IEA, 1971; and *The Energy 'Crisis' and British Coal*, Hobart Paper 59, IEA, 1974.

TABLE I

THE COAL INDUSTRY IN GREAT BRITAIN, 1947 TO 1980

	SUPPLY			GROSS CONSUMPTION		EMPLOYMENT	PRODUCTIVITY
	Production* Deep-mined	Opencast	Imports	Home	Exports and foreign bunkers	Number of wage-earners in collieries at year-end†	Output per man-year†
	(million tonnes)			(million tonnes)		(thousands)	(tonnes)
1947	190	10	1	188	5	717	269
1957	213	14	3	216	8	710	301
1967	168	7	—	167	2	385	416
1977	107	14	2	124	2	239	442
1980	112	16	7	124	4	228	483

*Including 'licensed' (non-NCB) mines but not recovered slurry.

†NCB mines only.

Sources: Ministry of Power Statistical Digests, Department of Energy Digests of UK Energy Statistics, Energy Trends (Department of Energy). Figures originally expressed in tons have been converted to metric tonnes.

[21]

TABLE II

INLAND CONSUMPTION OF COAL
IN THE UK, 1957 AND 1980

	1957		*1980*	
	million tonnes	*% of total*	*million tonnes*	*% of total*
Power stations	47.1	21.8	89.6	72.6
Coke ovens	31.2	14.4	11.6	9.4
Gas works	26.8	12.4	—	—
Industry	38.1	17.6	7.8	6.3
Domestic*	36.2	16.7	7.3	5.9
Railways	11.6	5.4	—	—
Other	25.3	11.7	7.2	5.8
TOTAL	216.3	100.0	123.5	100.0

*House coal and miners' coal.

Sources: Digest of UK Energy Statistics, 1976 and 1980; *Energy Trends* (Department of Energy), March 1981. Figures originally expressed in tons have been converted to tonnes.

fluctuations, particularly in coal consumption, which fell in 1978 but then rose by about 7½ per cent in 1979. Most likely the 1979 change was to a considerable extent a consequence of the very cold winter of 1978-79: in the first quarter of 1979 the mean air temperature in Britain was 2·2°C lower than the 1941-70 average.[1] In 1980, which was warmer than 1979, inland consumption of coal fell about 4½ per cent. At the same time as sales were declining in 1980, home coal production increased and there was a sharp rise in net imports, which approximately doubled; both the Central Electricity Generating Board and the British Steel Corporation found it cheaper to import foreign coal. Since home and imported supplies rose at a time of falling consumption, coal stocks mounted and by end-1980 were 10 million tonnes higher than a year earlier.[2]

In addition to the effects of warmer weather in 1980, the recession—and in particular the troubles of the British steel

[1] Temperature statistics are in Department of Energy, *Digest of UK Energy Statistics*, 1980, HMSO, Table 101.

[2] Statistics for 1980 can be found in Department of Energy, *Energy Trends*, March 1981.

industry—must also have affected British coal consumption adversely in the recent past. It is at present impossible to isolate the separate effects of these various influences. About all we can reasonably conclude is that the declining trend of coal consumption from 1957 until the mid-1970s has given way to a period in which consumption appears to be fluctuating around a more stable trend. Whether the change is temporary or permanent remains to be seen.

The considerable changes in the pattern of United Kingdom coal consumption in the last 20 years or so are illustrated in Table II. The industry has lost its old railway and gas markets completely: the railways changed to diesel and electric traction; the gas industry moved first to oil, as coal prices rose relatively to oil prices, and then to natural gas. Industrial and domestic sales have also contracted sharply. The NCB is now primarily a supplier of fuel to the electricity supply industry, which accounts for over 70 per cent of coal consumption, and to coke ovens which take about 9 per cent.

If coal markets in the 1970s (Table III) are examined in more detail, it is clear that power stations are not only now the principal outlet for coal; they have also for some years been the only expanding market coal has enjoyed. Between 1970 and 1980, sales to coke ovens, industry, the household market and miscellaneous users were more than halved (Table III). Sales to power stations crept up in the mid-1970s and then increased substantially in 1979, maintaining their higher figure in 1980.

Industrial relations and wages in a shrinking industry

Industrial relations in the coal industry are clearly of central importance. The NCB, which has a state-granted monopoly of coal production, is faced by a powerful industrial trade union (the NUM) and some smaller unions which appear to play only a minor rôle in negotiations. During the 1960s when oil prices were falling relatively to coal prices, and coal output and sales were therefore declining, the bargaining position of the mineworkers was weak and they accepted a considerable fall in the mining labour force and comparatively small pay increases. In the words of the 1972 Wilberforce Inquiry[1] into miners' pay:

'This rundown [of the labour force], which was brought about

[1] *Report of a Court of Inquiry*, Cmnd. 4903, HMSO, February 1972.

TABLE III

UNITED KINGDOM COAL CONSUMPTION BY MARKET, 1970 TO 1980

million tonnes

	1970	1973	1974	1975	1976	1977	1978	1979	1980
Electricity supply industry	77.2	76.8	67.0	74.6	77.8	80.0	80.6	88.8	89.6
Coke ovens	25.3	21.9	18.5	19.1	19.4	17.4	15.0	15.1	11.6
Industry	19.6	12.1	11.1	9.7	9.0	9.0	8.6	9.2	7.8
House coal*	18.2	12.7	12.0	9.9	9.5	9.6	8.7	8.9	7.3
Other	16.6	9.8	9.3	8.9	7.9	8.0	7.6	7.4	7.2
TOTAL	156.9	133.3	117.9	122.2	123.6	124.0	120.5	129.4	123.5

*House coal and miners' coal.

Source: *Digest of United Kingdom Energy Statistics*, 1980; *Energy Trends* (Department of Energy), March 1981.

[24]

TABLE IV

LABOUR COSTS IN NCB MINES, 1957 TO 1978-79

Wages and wage-related costs per tonne (£)*

1957	2.37	1968-69	2.21
1958	2.40	1969-70	2.30
1959	2.33	1970-71	2.78†
1960	2.31	1971-72	3.50
1961	2.36	1972-73	3.71
1962	2.24	1973-74	4.92
1963-64	2.21	1974-75	6.22
1964-65	2.25	1975-76	8.47
1965-66	2.29	1976-77	9.62
1966-67	2.37	1977-78	10.83
1967-68	2.31	1978-79	12.62

*Available only for NCB financial year (April-March) from 1963-64 onwards.

†The increase in 1970-71 is partly attributable to the inclusion of National Insurance and pensions excluded from figures for previous years.

Sources: Ministry of Power *Statistical Digests* and *Digests of UK Energy Statistics.*

with the co-operation of the miners and of their Union, is without parallel in British industry in terms of the social and economic costs it has inevitably entailed for the mining community as a whole.'

On earnings, Wilberforce pointed out that, although in 1960 the miners had been third in a table of earnings in 21 industries, by October 1970 they were 12th.

As oil prices began to rise in about 1970, the bargaining position of the mineworkers was naturally strengthened and they sought large increases in earnings. They banned overtime late in 1971 and went on strike in January 1972: on the recommendation of Wilberforce, they obtained a wage increase substantially higher than the NCB had said it could finance. A further overtime ban and strike followed in the winter of 1973-74, when oil supplies were short: a settlement was achieved after the return of a Labour Government in March 1974 with further substantial wage increases. The acceleration of earnings per tonne from the early 1970s onwards, after a long period of relative stability, is shown in Table IV.

TABLE V

PRICES OF COAL AND OIL USED BY INDUSTRY, 1967 TO 1980

	Coal	Oil	Coal price as %
	pence per therm		of oil price
1967	2.15	2.06	104
1968	2.08	2.25	92
1969	2.11	2.22	95
1970	2.53	2.22	114
1971	3.02	3.33	91
1972	3.25	3.18	102
1973	3.40	3.11	109
1974	3.70	7.37	50
1975	5.55	9.28	60
1976	6.87	10.63	65
1977	8.20	13.48	61
1978	8.90	12.64	70
1979	10.36	15.70	66
1980 2nd Quarter	12.5	22.5	56

Note: in 1974 there was a change in the method used to compile the information.

Sources: Digests of United Kingdom Energy Statistics, 1979 (Table 87) and 1980 (Table 89); *Energy Trends* (Department of Energy), December 1980, Supplementary Table.

These wage increases inevitably meant higher coal prices (about half the NCB's costs are wages and related charges), and there was also from the late 1960s an acceleration of general inflation which increased both labour and other costs. To illustrate the magnitude of the price increases, Table V compares the prices of coal and fuel oil sold to large industrial consumers from 1967 to mid-1980. From about 1970 coal prices began to rise considerably, after stability in the late 1960s. Oil prices had also started to increase, so that up to 1973 there was little change in the relative prices of the two fuels. Then the massive oil price rises of 1973 for a time raised oil prices sharply relatively to coal prices. From 1974 to 1978 coal prices rose much faster than oil prices before the price of oil again increased rapidly in

TABLE VI

UNITED KINGDOM PRIMARY ENERGY CONSUMPTION, 1950 TO 1980

	1950		1960		1970		1973		1978		1980	
	mtce	% of Total	mtce	% of Total	mtce	% of Total	mtce	% of Total	mtce	% of Total	mtce	% of Total
Coal	204	89.5	199	73.7	157	46.6	133	37.7	120	35.4	122	37.1
Oil	23	10.1	68	25.2	150	44.5	164	46.4	139	41.0	121	36.8
Natural Gas	—	—	—	—	18	5.3	44	12.5	65	19.2	71	21.6
Nuclear	—	—	1	0.4	10	3.0	10	2.8	13	3.8	13	4.0
Hydro	1	0.4	2	0.7	2	0.6	2	0.6	2	0.6	2	0.5
TOTAL	228	100.0	270	100.0	337	100.0	353	100.0	339	100.0	329	100.0

mtce = million tonnes coal equivalent.

Sources: Central Statistical Office, *Economic Trends,* Annual Supplement, 1979 Edn.; *Digest of UK Energy Statistics,* 1980; *Energy Trends* (Department of Energy), March 1981.

[27]

1979, restoring coal to about the same competitive position as in 1976 but less favourable than in 1974, 1975 and 1977. Then in 1980 the substantial oil price rise improved the competitive position of coal further, even though its price was also rising.

Finally, in this brief recent history of coal, Table VI illustrates how the share of coal in the UK energy market has changed. As its consumption fell in the 1950s and 1960s, essentially because its price was rising relatively to oil,[1] its share of the energy market dropped from about 90 per cent in 1950 to less than 47 per cent in 1970. There was a further decline to below 38 per cent in 1973. Since then its share has tended to stabilise as the relative price of oil has risen (Table V). From 1974 to 1980 the share of coal in the energy market varied between 35 and 37 per cent with no clear trend.

Government attitudes towards the coal industry

The coal industry is a very large employer (with nearly a quarter of a million employees), producing over one-third of home energy supplies and incidentally providing work in areas of high unemployment. Mining is an unpleasant and dangerous occupation, with a well-organised work-force having a strong sense of solidarity and a history of industrial and political struggle. For all these reasons the attitudes of politicians and civil servants towards the industry are of considerable importance.

Throughout the period of declining coal output and employment, the NCB and the NUM campaigned for government action to protect the industry. To some extent they succeeded —fuel oil was taxed and coal was not; the electricity supply industry was persuaded to burn more coal; coal imports were regulated; and substantial government financial assistance was given to the industry[2]—but they did not manage to persuade governments to agree to the coal output target of 200 million tonnes per annum the industry was urging from the late 1950s until the late 1960s. Government views of likely demand were indicated by the *Fuel Policy* White Paper of 1967,[3] which estimated coal consumption at 152 million tonnes in 1970 and 120 millions in 1975.

[1] *A Policy for Fuel?, op. cit.*, pp. 13-14.

[2] The forms of government aid to the coal industry are explained in *The Energy 'Crisis' and British Coal, op. cit.*, pp. 41-42.

[3] *Fuel Policy*, Cmnd. 3438, HMSO, November 1967.

TABLE VII

BRITISH DEEP-MINED COAL OUTPUT
ACCORDING TO NCB PLANS, 1985 TO 2000

	1985	1990	2000
		million tonnes	
Total, given no further major investment	120	105	80
Incremental output from unapproved projects at existing mines	—	9	15
Identified mines not yet approved	—	16	18
Other new mines not yet identified	—	—	37
	120	130	150

Source: Sir Derek Ezra, *Statement to Commission on Energy and the Environment*, 16 January 1979.

The oil price rises in the early 1970s apparently produced more sympathy for the industry from the government. The NCB's *Plan for Coal*,[1] published in 1974, which aimed to expand output to 135 million tonnes in 1985, was endorsed by the government; it was followed by *Coal for the Future* in 1977 which set out the Plan 2000 target of 150 million tonnes from deep mines and 20 million tonnes from open-cast operations in 2000. According to the NCB, deep-mined output would fall to only 80 million tonnes in 2000 if there were no further major investment schemes. Thus the Board estimated that 70 million tonnes of annual deep-mined capacity must be brought into operation between 1985 and 2000 to offset the exhaustion of existing pits and to raise annual output to 150 million tonnes. The NCB's views on how the 150 million tonnes might be provided from existing and new capacity are shown in Table VII.

The investment schemes required for the Plan 2000 target have not yet received government sanction. As is usual with the nationalised industries, individual schemes have to go to the responsible Ministry (here the Department of Energy), and

[1] National Coal Board, *Plan for Coal*, June 1974.

the more sensitive projects, such as the Belvoir proposal, then proceed to Public Inquiry. Plan 2000 was, of course, endorsed in general terms by government since it stemmed from the deliberations of the Tripartite Group (page 19), but providing the money is a different matter. Official written statements since *Coal for the Future* have expressed doubts about NCB production targets and have been lukewarm, if generally approving, about investment in the British coal industry, as shown by the following statements from the (Labour) Government's 1978 Green Paper on *Energy Policy*:[1]

'Up to 1985 the 135 million tonnes of output aimed at by *Plan for Coal* may represent an upper limit to capacity. Both *Plan for Coal* and its successor will have to go smoothly, if output at or near the Board's proposed target of 170 million tonnes in 2000 is to be achieved.' (para. 6.21)

'The NCB is aiming to produce 150 million tonnes of deep-mined coal in 2000 together with 20 million tonnes of open-cast. It is not possible to be sure how speedily new capacity can be introduced, and these objectives may not be fully achieved.' (para. 14.15)

'The production figures (for 2000), particularly for coal and nuclear, are upper limits and will not be achieved without very great efforts.' (para. 14.17)

'Using the appropriate financial tests, we should proceed with the creation of further new capacity in the coal industry, over and beyond the *Plan for Coal*, to come into production mainly in the late 1980s and 1990s . . .' (para. 14.27)

The qualification in the final quotation about 'appropriate financial tests' is particularly significant: only investment which meets the rate of return criterion laid down by the government of the day will be allowed to proceed. In other words, although the Labour Government's Green Paper blessed the NCB's expansion plans in general terms, the Board by no means had *carte blanche* to proceed. Under the terms of the 1978 White Paper,[2] a nationalised industry is supposed to earn a pre-tax real rate of return of 5 per cent on its new investment as a whole. Thus Labour's policy was evidently to allow the NCB to invest in new capacity provided it could show at least such a 5 per cent return. In effect, the policy set out in the Green Paper seemed to throw the investment decision back on to

[1] *Energy Policy: A Consultative Document*, Cmnd. 7101, HMSO, 1978.
[2] *The Nationalised Industries*, Cmnd. 7131, HMSO, 1978.

market forces. If coal costs were held down so that the NCB's competitive position improved, new investment would be relatively easy to justify. But if coal costs and prices seemed likely to rise in step with or faster than those of other fuels, it would become difficult for proposed investments to pass the 'appropriate financial tests'.

Manipulation of costs and optimism

It would be naïve to imagine that under Labour the market would have been allowed to rule without disturbance. When investment appraisals are made there is scope for considerable conscious or unconscious manipulation of the costs and revenues which form the basic data; at the least, some 'appraisal optimism' is built in by project champions. When negotiations on projects occur in private between the NCB and the Energy Department, there can be no certainty about just what influences determine the outcome. One of the advantages of a Public Inquiry is that it offers the opportunity to seek information about the relevant investment appraisals; if the applicant at the Inquiry is unwilling to provide such data, that in itself may be highly significant. At the Vale of Belvoir Inquiry neither the NCB nor the Department of Energy would supply the basic appraisal data for the three pits proposed; nor would they explain in detail the assumptions made, although the Department claimed that preliminary calculations showed the proposal had passed its financial tests.

Since the change of government in 1979, there seems to be a harder government attitude towards the coal industry. We must accept that political and social realities dictate that in public speeches Department of Energy Ministers must emphasise the 'vital rôle' coal will play in Britain's energy future. Such action as has been taken, however, has reduced the direct protection to British coal, already fairly low by the standards of other EEC countries. The Coal Industry Act of 1980[1] eliminates over a period the deficit grants (worth about £190 million in 1979-80) previously given to the Coal Board to make up for its losses; by 1983-84 the NCB is supposed to break even without such grants. 'Social' grants, covering, for example, payments

[1] The Third Reading Debate on the Coal Industry Bill gives the relevant statistics, especially the speech by Mr David Howell, *House of Commons Hansard*, 24 July 1980, cols. 813-820. 'Coal's tough break-even target', *Financial Times*, 4 August 1980, comments on the provisions of the Bill.

to the miners' pension scheme and worth about £60 million in 1979-80, will remain; and there will be a new provision to allow deferment of interest charges on some capital projects until they start to show a return. Apparently such deferment will provide £30-40 million a year to the NCB. On balance, if the Government stands by its policy, the Board will in the next few years receive significantly less subsidy than recently.

As well as examining political speeches and legislation passing through the House of Commons, we must observe the attitudes of civil servants towards the coal industry since they play a significant part in limiting Ministerial policy options and, directly or indirectly, in determining policy. Civil service attitudes as expressed in veiled terms in the 1978 Green Paper (page 30) were mildly favourable towards coal. An unusual feature of the Belvoir Inquiry was that the Department of Energy allowed one of its senior officials to be cross-examined about its views on coal. It was thereby undoubtedly placed in an unfamiliar and rather delicate situation, rendered even more awkward by the Coal Board's attitude towards the Inquiry. Incredible as it may seem, the Board decided it was unnecessary to provide the Inquiry with up-to-date coal demand forecasts of its own, even though it was proposing a major expansion of coal capacity: at their peak, the three proposed mines in North East Leicestershire would produce about 7 million tonnes a year. Instead, the NCB said it would rely on the Department of Energy's forecasts as set out in *Energy Projections 1979*[1] and a document produced at the Inquiry entitled *Assessment of Energy Requirements*.[2] According to the Board it had examined the Department's projections in detail and had decided to accept them as they stood. Thus the Energy Department was cast in the uncomfortable rôle of defender of the Coal Board's expansion plans.

Energy Projections 1979 estimated coal consumption in the year 2000 at 128 to 165 million tonnes and coal production at 137 to 155 million tonnes. Although we should not take such projections too seriously, an obvious difficulty for the Department of Energy was to explain why even the top ends of its forecast ranges fell below the NCB's target of 170 million tonnes.

[1] Department of Energy, *Energy Projections 1979*, 1979.

[2] Department of Energy, *Assessment of Energy Requirements*, 1979 (Note to the Vale of Belvoir Inquiry).

More important was the peculiar concept of an import 'gap' used by the Department of Energy to justify new coal investment projects. The Department's view, as expressed in *Assessment of Energy Requirements*, appeared to be that 'failure to develop our indigenous coal resources efficiently' would lead by the end of the century to a gap between fuel demand and home fuel supplies which would have to be filled by imports. One could scarcely quarrel with the sentiment that coal resources should be developed 'efficiently', but the Department's *Assessment* gives little help in determining what would constitute 'efficient' development. Instead it seems to believe in import minimisation *per se* as a proper aim of fuel policy. We discuss in more detail this import gap concept in Section VI.

II. COAL IN THE WORLD ENERGY MARKET

Recent trends and prospects

One of the consequences of the energy 'crisis' of the mid- and late-1970s has been renewed interest in coal as a source of energy. Rising oil prices have stimulated the search for alternatives to OPEC oil, whether nuclear power, 'income' sources[1] of energy such as solar, wind and tidal power, or old-established fuels such as coal.

Coal consumption and production almost halved in Britain between the late 1950s and 1979 (Table I), whilst the share of coal in the energy market dropped from about 90 per cent at the end of the Second World War to about 37 per cent in 1980 (Table VI). In the world as a whole, its share in 1950 (about 60 per cent) was considerably lower than in Britain, but it then fell much less during the post-war period: by 1979 it was about 32 per cent—only a little less than in Britain. Table VIII illustrates the trends in world energy consumption from 1950 to 1979. In Britain coal sales dropped substantially, but the tonnage consumed in the world as a whole rose considerably, almost doubling between 1950 and 1979—a rate of increase of just over 2 per cent a year. Although European coal industries were declining because they were uncompetitive relatively to oil, coal fared much better where geological conditions were more favourable.

There was a marked change in energy trends after 1973 compared with the earlier post-war period (Table VIII). Relative price variations have had powerful effects, as indeed they had in the 1950s and 1960s. As fuel prices have risen and the world economy has grown more slowly, the rate of increase of world energy consumption has halved—from over 5 per cent a year between 1950 and 1973 to about $2\frac{1}{2}$ per cent since 1973. As would be expected the change in trend has been largest in oil, the price of which has risen most: the rate of growth of world oil consumption was less than 2 per cent a year between 1973 and 1979 compared with $7\frac{1}{2}$ per cent in the earlier post-

---

[1] As distinct from 'capital' sources such as coal and oil where production depletes a finite stack.

TABLE VIII

WORLD CONSUMPTION OF COMMERCIAL ENERGY, 1950 TO 1979

	1950		1960		1970		1973		1979		Av. annual compound rates of increase	
	mtce*	% of total	mtce	% of total	mtce	% of total	mtce	% of total	mtce	% of total	1950–73 %	1973–79 %
Solid fuels	1,534	61	2,206	52	2,418	35	2,503	32	2,965	32	2.2	2.9
Liquid fuels	672	27	1,358	32	2,936	43	3,598	45	4,020	44	7.6	1.9
Natural gas	244	10	594	14	1,368	20	1,623	21	1,945	21	8.6	3.1
Hydro and Nuclear electricity	42	2	85	2	154	2	186	2	280	3	6.7	7.1
TOTAL	2,492	100	4,243	100	6,876	100	7,910	100	9,210	100	5.2	2.6

*mtce = million tonnes coal equivalent.

Sources: 1950-73: *World Energy Supplies 1950-74*, United Nations, 1976. 1979: estimated from *BP Statistical Review of the World Oil Industry*, 1979.

[35]

war period. Incomplete statistics for 1980 suggest that there was a substantial fall in world oil consumption.[1] Consumption of natural gas, the price of which has also risen considerably, has been growing at about 3 per cent a year since 1973, compared with 8½ per cent earlier. Nuclear and hydro power are still relatively minor components of total world energy consumption,[2] although consumption of nuclear electricity rose quite rapidly (at over 20 per cent a year) between 1973 and 1979.

Revival in coal's competitiveness

Coal has clearly improved its competitive position in many countries since 1973 as oil and gas prices have risen. Consequently world coal consumption has increased at nearly 3 per cent a year instead of just over 2 per cent in the earlier post-war years, and the share of coal in world energy has remained at around 32 per cent (Table VIII).

It seems very likely that the energy trends of recent years will continue until the end of the century because of changes in relative fuel prices, and that there will be a revival of coal in the sense that its share of the world energy market will probably rise. Most forecasts agree about this anticipated revival,[3] though the record of energy forecasters does not inspire much confidence in their predictions, and the quantities now forecast for the late 20th century should be treated with considerable scepticism. In general terms, however, there is no reason to differ from the optimistic conclusions about the global growth prospects for coal in the more reputable studies.[4] Economically recoverable coal reserves are certainly plentiful enough to support rising consumption for many years and there is scope for a big increase in international trade in steam coal.

[1] World oil production fell slightly in the first half of 1980. ('World crude output falls in first half', *Financial Times*, 29 August 1980.)

[2] The conversion methods into coal equivalent used by the United Nations give nuclear and hydro electricity smaller shares of the energy market than would UK conversion methods.

[3] For example, OECD, *Steam Coal: Prospects to 2000*, 1978; World Coal Study, *Coal—Bridge to the Future*, Ballinger, 1980, and *Future Coal Prospects: Country and Regional Assessments*, Ballinger, 1980; and International Energy Agency, *Report of the IEA Coal Industry Advisory Board*, December 1980.

[4] Those mentioned in footnote 3 above and Exxon's *World Energy Outlook*, December 1979, especially pp. 16-17.

Nevertheless, it would be quite wrong to conclude that, because global prospects for coal appear bright, the industry will expand in every country. There are wide differences in geological, economic and social conditions among coal-producing countries which make for significant differences in their competitive strengths relatively to other fuels.

In particular, it is erroneous to assume that, because there is likely to be an expansion of coal production in such countries as the USA, Australia and South Africa where comparatively low-cost strip-mining techniques can be used,[1] there will also necessarily be increased output from the relatively labour-intensive deep mines of Western Europe. For several reasons (pp. 77-84), movements in fuel costs and prices will probably induce substitution of high-priced oil in Europe by coal imported from countries with low costs of production. In the 1960s, as relative prices moved in favour of oil, it was substituted for coal; that process is now likely to be reversed, but, unless governments impose import barriers, the main instrument of substitution will most probably be imported rather than indigenous coal. Thus a growth in *total* world coal output is quite consistent with stability or decline in the industries of Western Europe where coal will probably be supplanted partly by imports or by other forms of energy. In the post-war period up to 1973 world coal production was growing at a time when West European coal production was fast contracting.

World coal trends and the British coal industry

This simple proposition about *differences* between coal industries is emphasised because of the widespread tendency in Britain to confuse worldwide coal prospects with the outlook for British coal. This confusion is compounded, wittingly or not, by the Coal Board's public statements which emphasise the prospect of global oil and gas shortages and a consequent revival of coal whilst saying very little in specific terms to justify its expansion plans for British coal. As an example, we quote in full the only reference (paragraph 2) to coal demand in the NCB's 'Statement' accompanying its application to mine coal in North East Leicestershire.

'Because the mines will not reach full production until after 1990 it is from a long-term point of view that the need for them must

---

[1] Prospects in the major coal-producing and -exporting countries are discussed in sections I and L of *Steam Coal: Prospects to 2000, op. cit.*

be assessed. As an extractive industry, coal-mining needs new capacity merely to maintain production. The Board consider that this need will be the greater as the potential demand for home-produced coal will be on a rising trend from the mid-1980s for a number of reasons. Demand for energy is expected to continue to rise, in spite of the effects of energy conservation. The world price of oil is expected to rise significantly and by 2000 may have doubled in real terms. As a consequence, there will be an increase in the value of North Sea oil and natural gas, and both will tend increasingly to be reserved for premium uses. It is unlikely that by the end of the century nuclear power or "renewable" sources of energy will have expanded sufficiently both to meet increased demand and to take the place of oil and natural gas in the non-premium markets. The Board consider it will be highly disadvantageous to meet the rising potential market for coal by imports, since the price of imported coal will be greatly affected by the rising world price of oil, so that quite apart from strategic issues large coal imports would represent a considerable and rising burden on the balance of payments. The potential increase in the demand for coal should therefore be met by higher coal production in the United Kingdom. This would have the added advantage that it provided employment at home.'

What is meant by 'potential demand' for home-produced coal is unclear. The paragraph is a curious mixture of statements about world energy trends and comments about fuels other than coal. Though it is apparently about demand, the approach is essentially from the supply side. Demand is somehow determined—it is 'expected to continue to rise'—and the question then is how to find supplies of fuels to satisfy it. Prices are mentioned only for oil (the world price only) and for imported coal. There is no mention whatsoever of cost and price trends for British coal nor of the place of consumers' wants in the energy market. At the end of the paragraph there is a rather clumsy attempt to attack the case for importing coal and a remarkable *non sequitur* that the 'potential increase in demand should *therefore* be met by higher coal production in the United Kingdom'.

A single paragraph from a Planning Application hardly justifies lengthy attention. More important, the Board's case at the Belvoir Inquiry appeared to us to be related essentially to its desire to *supply* more coal. Evidently its plans are predominantly supply-determined, aiming to achieve production targets it has set without specific analysis of likely consumer demand. A production-oriented attitude with relatively little regard for the consumer interest is not uncommon in large organisations which enjoy semi-monopoly power and substantial support from government. However, we are not here concerned primarily with what determines the NCB's view of the future, but with attempting to examine in some detail whether consumer demand is likely to be sufficient to justify the Board's output programme. We return below to the world coal market and, in particular, to the prospects for importing coal into Britain.

III. THE ENERGY 'CRISIS', BRITISH COAL, AND TOTAL FUEL DEMAND

British coal and the energy 'crisis'

The energy 'crisis' of the early 1970s was the event which stimulated the NCB into laying its present expansion plans. It now attempts to justify them by arguing that shortages of oil and other fuels will generate a 'need' for coal, although neither in its publications nor at the Belvoir Inquiry has the Board presented evidence to show that there is likely to be a demand from consumers for the coal it intends to produce. In economic terms the Board's case appears to be that there will be a considerable increase in total fuel demand and that rising prices of oil and other fuels will shift relative fuel prices in favour of coal. In addition, the Board claims there are security advantages attached to British coal (discussed in Section VI).

The case for coal expansion based on rising fuel prices is more complicated than it might appear at first sight, so we examine it in some detail here and in Sections IV and V. Essentially, there are two aspects (apart from changes in government policy), each of which will affect the British coal industry very differently. If it is true that fuel prices will rise substantially in real terms, the two effects will be:

1. The *total demand for energy* in Britain will expand less than it would otherwise have done if average fuel prices rise relatively to other prices, both because the rate of economic growth is likely to decline and because greater efficiency in fuel use will be stimulated. Other things equal, the depressing effect on total energy demand will tend to *reduce* the demand for coal, which will be competing within a smaller market than if fuel price increases had been lower.

2. If the *prices* of particular fuels increase at different rates, relative fuel prices will change and their *market shares* will therefore alter. *If* it could be assumed that the relative price of British coal will fall, it would be reasonable to expect the market share of coal to increase. Whether such an increase would be sufficient to offset the reduction in the

[40]

demand for coal resulting from lower total energy demand
(1) cannot be deduced *a priori*.

The impact of rising fuel prices on British coal consumption
is thus more complex than might appear from the NCB's
statements, which create the impression that such movements
are necessarily consistent with an expanding demand for coal.
To analyse in logical order the outlook for coal consumption,
we first examine the prospects for total energy demand, and
then (Sections IV and V) investigate the future competitive
position of British coal which will determine its share of that
total demand, on the assumption that government policy
remains unchanged. Appropriate government policies are discussed later (Section VI).

Fuel demand in Britain

There are numerous ways of aggregating quantities of various
fuels to arrive at estimates of total fuel demand. None of them
is particularly satisfactory but here we use the 'coal equivalent'
measure, since our prime concern is with the coal industry and
there are coal equivalent statistics for the UK which go back
for many years.[1]

On the coal equivalent measure, UK energy demand has
increased fairly slowly in comparison with other countries. In
the period 1950 to 1973, for example, it rose at a compound
rate of about 1·9 per cent a year, equivalent to roughly two-
thirds the rate of growth of real gross domestic product (GDP)
over the same period (2·8 per cent a year). In most other in-
dustrialised countries energy consumption grew faster, partly
because of more rapid GDP expansion and partly because the
rate of increase of energy demand per unit increase in GDP
was higher.

Following the increases in oil and other fuel prices during
1973-75 and the recession of that period, total UK energy
consumption fell (Table VI). In 1978 primary energy consumption was still about 4 per cent lower than in 1973, although real GDP had grown about 4½ per cent. In other words,
the amount of primary energy consumed per unit of real GDP

---

[1] Statistics for recent years are in the Department of Energy's *Digest of Energy
Statistics 1980* and its predecessors. A useful source of historical statistics is *The
British Economy: Key Statistics*, Times Newspapers for London and Cambridge
Economic Service, 1971.

[41]

fell by more than 8 per cent during the five years. Energy trends since 1978 are difficult to interpret because of the recession, big temperature variations between 1978, 1979 and 1980, and the build up of fuel stocks, both voluntarily in expectation of oil price rises and involuntarily because of falling demand in 1980. Primary energy consumption increased by 4·7 per cent in 1979 to approximately what it had been in 1973; in 1980, however, it appears to have fallen by about $7\frac{1}{2}$ per cent to well below the 1973 peak.

Forecasts of energy demand

In looking ahead to likely fuel demand in the later years of this century, we must proceed with a proper appreciation of the enormous uncertainties in such forecasting. At the time of the Belvoir Inquiry we produced fuel demand forecasts, based on a sector-by-sector analysis of the energy market, which were intended to give a very approximate indication of trends in demand and to provide a check on the Department of Energy's estimates as set out in *Energy Projections 1979* (details of our estimates are in Table IX). We discuss below energy demand forecasting in general, NCB and Energy Department projections, and the main features of our own analysis.

At one time energy 'forecasts' were often produced very crudely as projections of time-series or as simple relationships between energy consumption and real GDP. Such naïve methods may seem to have worked well in comparatively settled periods like the 1960s, but they are devoid of economic content—in particular *they ignore relative price effects*. The events of recent years have induced most energy forecasters to attempt to incorporate in their models the energy-saving effects of rising fuel prices.[1]

One surprising revelation at the Belvoir Inquiry was that the NCB had not published a total primary fuel demand forecast[2] for the UK since *Coal for the Future* in 1977 and did not produce one for the Inquiry. *Coal for the Future* set the target for

[1] An attempt to calculate income and price elasticities is in George Kouris and Colin Robinson, 'EEC demand for imported crude oil, 1956-85', *Energy Policy*, June 1977.

[2] The forecasts of the NCB and of the Department of Energy relate to total primary *fuel*, not total primary *energy* as shown in Table VI. On UK definitions, primary fuel exceeds primary energy because it includes non-energy products and international bunkers. The difference between the two measures in recent years is shown in Table X.

TABLE IX

ESTIMATES OF UNITED KINGDOM
ENERGY CONSUMPTION BY FINAL USERS
IN THE YEAR 2000

	1977* Actual	2000 Department of Energy Estimates	Authors' Estimates
		thousand million therms	
Domestic	15.0	15.2–15.8	14.3–15.5
Iron & Steel	4.9	5.0– 6.0	4.4– 4.8
Other Industry	16.4	21.0–24.1	19.0–21.5
Transport	13.0	15.5–18.7	15.0–17.0
Other Consumers	7.6	8.6–10.5	8.0– 9.1
	56.9	65.3–75.1	60.7–67.9

*1977 is used as the base year in this Table because it is the base for *Energy Projections 1979,* the source of the Department of Energy statistics.

TABLE X

PRIMARY FUEL AND PRIMARY ENERGY
CONSUMPTION IN THE UNITED KINGDOM,
1973 TO 1979

	Total Primary Energy	*Total Primary Fuel*
	million tonnes coal equivalent	
1973	353.5	382.6
1974	337.5	365.8
1975	324.8	347.0
1976	329.8	353.1
1977	338.4	359.7
1978	339.8	360.3
1979	355.9	376.6

Note: Primary energy includes products used for energy purposes. Primary fuel includes, in addition, non-energy oil products (for example, chemical feedstock and lubricants) and international bunkers.

Source: Digest of UK Energy Statistics, 1980.

[43]

coal output at 170 million tonnes in 2000, a figure on which the NCB's expansion programme is still based. It also stated that total primary fuel demand in the UK in 2000 would be in the range 500 to 650 million tonnes coal equivalent (mtce) 'or more'. Presumably, therefore, the figure of 170 million tonnes coal output was consistent with the figure of at least 500-650 mtce total fuel demand, giving coal a fuel market share of 26 to 34 per cent.

In the last few years virtually all fuel forecasts have been substantially reduced. For example, the Energy Department, which in 1977 had the same forecast[1] as in *Coal for the Future* except for the 'or more' qualification made by the NCB, has subsequently revised its estimates downwards on two occasions —in the 1978 Green Paper on *Energy Policy* to between 450 and 560 mtce, and in *Energy Projections 1979* to between 445 and 510 mtce. The extremities of the Department's forecast range are now, respectively, 11 per cent and 22 per cent lower than the 1977 forecast. This tendency to reduce energy forecasts is by no means a peculiarly British phenomenon. International oil companies have all reduced their forecasts significantly. The EEC's energy demand forecast for 1985 is now 27 per cent lower than its forecast in 1973,[2] and US energy forecasts have fallen sharply.[3] Over the last few years, forecasters have gradually been coming to terms with poorer prospects for economic growth and with the effects on fuel efficiency of rising energy prices.

The one forecast of total fuel demand which does not appear to have been reduced is the NCB's—or, at least, if it has been reduced there is no explanation of why the 170 million tonne coal output target has not also been changed. If the NCB's total fuel demand forecast is too high there is a strong implication that the figure of 170 million tonnes—with which it was presumably consistent initially—is also too high. It is implausible to assume that the relationships between coal demand and its determinants, or those determinants themselves, have so changed that the 170 million tonne estimate remains correct despite the reduction in total energy forecasts. If, for example,

---

[1] In Department of Energy, *Energy Policy Review*, Energy Paper No. 22, 1977.

[2] Commission of the European Communities, *The Energy Programme of the European Communities*, COM(79)527, 4 October 1979, para. 18.

[3] For example, Amory Lovins, 'Safe Energy', *Resurgence*, January/February 1979, pp. 20-21.

the Energy Department's latest total fuel forecast for the year 2000 of 445-510 mtce is accepted, the 170 million tonne figure would give coal a market share of 33 to 38 per cent, compared with the share projected in *Coal for the Future* of 26 to 34 per cent. Does the Coal Board really believe its market share will grow so much more than it thought three years ago? Or has it simply failed to adjust downwards its coal output and sales estimates in line with changing views of the energy market? In that event, its plan to produce 170 million tonnes in the year 2000 is no more than a relic of a now outdated forecast.

Oddities and inconsistencies at Belvoir

In using the Energy Department's total fuel forecasts by way of example we do not wish to suggest that much weight should be placed on them. Indeed, one of the features of the Belvoir Inquiry was the numerous inconsistencies and oddities which came to light when the Department's witness was cross-examined about its forecasts. To the extent that these curious features relate to the competitive position of coal, they are discussed in Sections IV and V. But some of them concern the total fuel forecasts.

Energy Projections 1979 (Introduction, paragraph 2), for example, states that, to project fuel demand to the end of the century, alternative real GDP growth rates of 2 per cent and 3 per cent a year were used by the Energy Department. In cross-examination, it turned out that the precise higher assumption used by the Department was 2·7 per cent a year. If the Department wanted to round off its growth rates, most people would have expected the higher assumption to be stated as 2½ per cent.

The difference in practice is no more than irritating and misleading to anyone who wants to check the Department's forecasts. As an example of more serious inconsistencies, we may note that the Department assumes world crude oil prices will rise about 2½ times between the late 1970s and 2000, and yet takes as its upper estimate of real GDP growth a figure (2·7 per cent a year) approximately the same as that achieved between 1950 and 1973 when oil prices were stable and much lower than now in real terms. The Department apparently believes, in the face of recent experience, that economic growth is unresponsive to changes in the real price of oil since, for both of its GDP growth cases, it uses the same assumptions about oil price increases. Another oddity is the Department's estimate (para. 31 of *Energy Projections 1979*) that, beyond the year 2000, the rate of increase of total energy consumption will accelerate. Since no details of economic growth or price assumptions for this period are given, the Department's witness was questioned at the Belvoir Inquiry: the only answer which could be elicited was the question-begging response in a subsequent letter that energy supplies in this period would come increasingly from electricity and synthetic natural gas which would make primary fuel use less efficient.

A detailed examination of the Department of Energy's forecast of energy demand by sector reveals further strange assumptions which tend to bias the forecasts upwards. Even in mid-1979 when *Energy Projections 1979* was published, for example, it was surprising to read (para. 6, Annex 1) that the energy forecasts for the iron and steel sector assumed British steel output would rise to as much as 26 to 30 million tonnes by 2000 from 20 millions in 1978. The Department's witness at the Inquiry explained after questioning that the Department of Industry was making a different forecast, assuming steel output would be 21 to 25 million tonnes in 2000—which may still be on the high side.

Incomprehensible assumptions about manufacturing output

The most curious feature of the sector forecasts, however, relates to fuel consumption by industry other than iron and steel. We were unable to understand how the Energy Department could forecast a growth in consumption by this sector as high as 1·1 to 1·7 per cent a year (para. 7, Annex 1 of *Energy Projections 1979*), given its assumptions about GDP growth and energy conservation. The Department eventually explained in a letter to the Inquiry Secretariat that a key assumption was for British manufacturing output to increase as fast as real GDP up to the end of the century. This assumption is so much at variance with recent experience—between 1973 and 1979 manufacturing output *fell* by over 4 per cent whereas real GDP *increased* by about 5½ per cent—that it seems incomprehensible. There was no attempt to justify such a radical departure from recent trends. British manufacturing industry may well recover, but the *scale* of recovery assumed by the Energy Department seems unlikely, to say the least. Since 'other industry' is the largest single sector of energy consumption in

Britain, this idiosyncratic assumption imparts a significant upward bias to the Energy Department's forecasts compared with what most students of the British economy and energy market might expect. Incidentally, it is also the main reason for the conclusion of both the Energy Department and the NCB that a substantial growth in industrial fuel consumption will open up a big new market for coal (Section V).

The conclusions of our own estimates of the total demand for energy are summarised in Table IX. They were intended mainly as a check on Department of Energy projections. The real GDP growth assumptions we used (2 and $2\frac{1}{2}$ per cent a year) do not differ much from the Energy Department's assumptions (2 and 2·7 per cent). But our pattern of economic growth up to 2000 was different, and we attempted to be consistent in our assumptions. We assumed that real energy prices would double in the higher GDP growth case and be multiplied $2\frac{1}{2}$ times in the lower growth case. The other main assumptions were that real personal disposable income would increase at 2 to $2\frac{1}{2}$ per cent a year; that crude steel output in the year 2000 would be in the range of 20 to 22 million tonnes; and that manufacturing output would rise at 1 to $1\frac{1}{2}$ per cent a year. In the light of recent British economic performance, all these growth assumptions might be criticised on the ground that they are optimistic. For that reason, our fuel consumption estimates for 2000 may be on the high side.

Our conclusion is that the total demand for primary fuel in the year 2000 may be in the range of 400 to 450 mtce, compared with 377 mtce in 1979 and with the Energy Department's projection for 2000 of 445 to 510 mtce.[1] We do not suggest that our estimates should be taken too seriously, since they relate to a period many years ahead. But we do believe there is evidence that the Department of Energy's estimates of total fuel consumption are still too high. It would hardly be surprising if further reductions in the Department's projections were made. Certainly we would hope to see more consistency of assumptions, more explicit comment about the nature of those assumptions and, in particular, a more realistic assessment of the prospects for fuel consumption by British manufacturing industry.

---

[1] To obtain these coal equivalent primary fuel estimates, the estimates in Table IX have to be converted from therms (using Department of Energy conversion factors) and additions made for non-energy uses of oil, international bunkers and consumption of fuel in the energy industries.

IV. PROSPECTS FOR BRITISH COAL: WILL IT BE COMPETITIVE?

The British fuel market is thus likely to expand relatively slowly to the end of the century. What of the future competitive situation of coal, which will be the main influence on its share in the total market? This question leads into controversial matters since we have to investigate the (future) trend of costs (including wages) and prices in mining.

Market trends in the recent past

The recent history of coal consumption was discussed in Section I. To summarise, Table VI showed that the share of coal in the UK energy market fell sharply during the post-war period until the mid-1970s when, as relative prices moved in favour of coal (Table V), its market share stabilised and consumption fluctuated within the range of 120 to 130 million tonnes a year. It is clear from Table III, however, that the comparative stability of coal consumption in recent years is a consequence of two conflicting (and approximately offsetting) trends—a rise in sales to power stations and a continued fall in sales to all other major markets.

The evidence of the last few years illustrates the difficulties the NCB will have in persuading consumers to buy the considerable extra quantities of coal it wants to produce. Despite the substantial price advantage coal has enjoyed over oil (Table V), it has increased its sales and market share only in the large integrated system of the electricity supply industry, where usage of coal-fired and oil-fired power stations can readily be varied in response to relative price changes.

In the smaller sectors of consumption, where sales have continued to decline, coal may have been adversely affected by special factors in recent years. For example, the troubles of the steel industry have reduced NCB sales to coke ovens. Whether that is any real comfort for the Coal Board, however, is doubtful since the British steel industry's troubles may well be persistent and imports of coking coal have been undercutting

[48]

TABLE XI

FINAL FUEL CONSUMPTION BY INDUSTRIES OTHER THAN IRON AND STEEL, 1973 AND 1979

	1973		1979	
	Million therms	*% of Total*	*Million therms*	*% of Total*
Coal, coke and coal-derived fuels	3,252	16.9	2,485	13.6
Petroleum*	9,341	48.7	7,411	40.6
Natural gas, town gas and coke oven gas	4,247	22.1	5,811	31.9
Electricity	2,355	12.3	2,528	13.9
	19,195	100.0	18,235	100.0

*including a very small quantity of creosote/pitch mixtures.

Source: Digest of UK Energy Statistics, 1980.

NCB prices and increasing their share of the steel market (Section VI).

In the industrial market, coal sales have been affected by the period of recession and by the generally slow economic growth since 1973. Nevertheless, coal has fared badly in competition with other fuels. Total energy consumption by industries other than iron and steel declined by about 5 per cent between 1973 and 1979 (Table XI). Within that falling total, gas increased its share considerably and there was also an increase in the share of electricity. The market shares of both coal and oil have dropped since 1973 but, surprisingly, coal consumption has fallen by nearly 24 per cent whereas oil consumption is down by less than 21 per cent. The industrial market is one in which the Coal Board has great hopes of increasing its sales very considerably, partly because it seems to share the optimism of the Department of Energy about a recovery of British manufacturing (Section V). The recent trend of industrial coal sales must, therefore, be worrying for the NCB[1] and is indeed

[1] At the time of writing, insufficient information was available to include 1980 figures in Table XI. However, coal sales to industry fell by about 15 per cent in 1980 (Table III).

[49]

difficult to explain given the price advantage of coal. Possible reasons are the time-lag involved in switching from oil to coal (because investment in new plant or modification of existing plant is required) and the greater convenience in use of oil and gas. The 1978 Green Paper[1] suggested that the price of coal has to be up to 4 pence per therm lower in small- and medium-sized industry if it is to be competitive with other fuels. It is also possible that industrialists are sceptical whether the price advantage of coal will persist in the long run if imports are excluded. Furthermore, they may have fears of an interruption of British coal supplies like that of 1973-74—about which Sir (then Mr) Derek Ezra is reported to have said: 'Our customers will not accept our arguments about security of supply if we go on in this way.'[2]

A general conclusion that might be drawn from recent experience is that the NCB will have to keep coal prices significantly below oil prices, and convince its customers that it will be able to sustain its favourable competitive position, if it is to have any hope of raising its sales significantly. *Coal for the Future* does indeed mention the need for a 'substantial and sustained competitive margin' for coal.[3] Consumers will also have to be convinced that they are less likely to be subjected to interruptions in British coal supplies from strikes or other forms of industrial action than interruptions in oil supplies.

What determines the competitive position of coal?

The Coal Board appears to be relying on sharp rises in the price of other fuels to keep its competitive position favourable. At the Belvoir Inquiry it claimed to foresee big increases in the prices of oil, natural gas and imported coal on the basis of world energy trends, but it would give no specific views on what might happen to the price of its own product. Yet if British coal is influenced by these same energy trends its price will also rise, and it is by no means obvious that it will have the 'substantial and sustained' competitive margin on which NCB plans are based. Although no-one can hope to foretell accurately, in absolute terms, what will happen to fuel prices, simple economic analysis should be able to help us make a

[1] Cmnd. 7101, para. 613.

[2] *Financial Times*, 4 December 1973.

[3] *Coal for the Future, op. cit.*, para. 45.

[50]

TABLE XII

COSTS AND PROCEEDS PER TONNE:
NCB COLLIERIES, 1971-72 AND 1978-79

| | *1971-72* | | *1978-79* | |
	£	*% of total costs*	*£*	*% of total costs*
Wages and related costs	3.50	45.4	12.62	50.7
Materials and repairs	1.69	21.9	5.61	22.6
Power, heat and light	0.34	4.4	1.09	4.4
Other costs (including depreciation)	2.18	28.3	5.55	22.3
TOTAL COSTS	7.71	100.0	24.87	100.0
Total proceeds	6.50	84.3	24.48	98.4
Loss before charging interest	1.21	15.7	0.39	1.6

Source: Department of Energy, *Digest of Energy Statistics 1980,* Table 24.

broad assessment of how, *relatively to other fuels,* British coal prices are likely to vary.

An important feature of British coal-mining is its labour intensity. With present and foreseeable technology and British geological conditions, the bulk of coal mined is likely to come from deep-mining operations. Eventually, there might be a radical reduction in labour intensity (for example, through the development of an economic process to gasify coal *in situ*)[1] if the mining unions will accept it. But at least for the rest of this century and the early part of next, British coal is likely to be labour-intensive and relatively high-cost by world standards.

[1] 'NCB studies plans for automated mining', *Financial Times,* 7 July 1980.

[51]

The 1978 Green Paper[1] estimated the average pithead cost of UK coal at more than double the extraction cost of strip-mined coal in the US and Australia.

At present, labour costs account for over 50 per cent of total NCB colliery costs (Table XII). During the 1960s when wages were rising only slowly and productivity was increasing (Table I), labour costs per tonne were fairly steady (Table IV), falling to a low point of 45 per cent of total NCB costs in 1971-72 (Table XII). Subsequent big wage increases without productivity gains, however, pushed labour costs back to over 50 per cent of total costs so that the industry has not succeeded in reducing the labour intensity of its operations. The reduction in the losses on NCB colliery operations (Table XII) is essentially a result of big price increases during the period when oil prices were rising. By 1978-79 wage costs per tonne were 3·6 times higher than in 1971-72 and total costs per tonne were 3·2 times higher. Total proceeds per tonne (that is, the average pithead price of coal) grew faster—in 1978-79 they were 3·8 times their 1971-72 level. Coal prices will to a large extent be a function of relative trends in coal productivity and of earnings in the industry.

Productivity and earnings

If there is to be any prospect of a sustained improvement in the competitive position of British coal, considerable increases in productivity are a necessary (though not a sufficient) condition. If, to take an extreme example, productivity remained as at present, the planned increase in deep-mined output of nearly 40 per cent to the year 2000 would require about 90,000 additional miners. To recruit such large numbers into mining would almost certainly mean huge increases in mining wages which would very likely price coal out of the market, even if oil prices rose considerably. In practice, some productivity gains are likely. *Plan for Coal* assumed that output per manshift could be increased by at least 4 per cent a year up to 1985,[2] but this target is now far out of reach given the disappointing trend of the last few years. *Coal for the Future* was much more cautious about future productivity.[3]

[1] Cmnd. 7101, para. 3.25.

[2] *Plan for Coal, op. cit.*, p.14.

[3] *Coal for the Future, op. cit.*, para. 22.

TABLE XIII

PRODUCTIVITY AT NCB COLLIERIES, 1968 TO 1980

	Output per man-year	*Tonnes* *Overall output per man-shift*
1968	454	2.12
1969	468	2.21
1970	468	2.24
1971	478	2.23
1972	402	2.22
1973	464	2.29
1974	405	2.18
1975	471	2.28
1976	452	2.23
1977	442	2.18
1978	452	2.25
1979	462	2.25
1980*	483	2.28

*Partly estimated.

Sources: Digest of UK Energy Statistics, 1980, Table 25; Department of Energy, *Energy Trends,* January 1981.

Naturally enough, the NCB's public comment on trends in productivity concentrates on the relatively favourable experience of the recent past. Sir Derek Ezra is reported to have told miners in August 1980, for example, that

'the Board could continue to build on the success of 1979, when output was up 4 million tonnes, coalface productivity was at a record level and attendance had greatly improved'.[1]

Without wishing to detract from recent achievements in the coal industry, in the interests of accuracy it must be pointed out that the productivity increase is really a recovery from a depressed period in the mid- and late-1970s. Only now is productivity returning to what it was in the early 1970s. In 1980 output per man-year in NCB collieries was only slightly above its previous peak in 1971 and total output per man-shift (OMS) appears to have been about the same as in 1973[2] (Table XIII).

[1] 'Coal demand to dip by 5m tonnes', *Financial Times*, 20 August 1980.

[2] Accurate comparison with earlier figures is not possible since the basis of the output per man-shift statistics has been revised recently.

A decade with no improvement in productivity is a far cry from the 4 per cent a year rise assumed in *Plan for Coal*.

The NCB claims that the new pits it proposes will operate at above present average productivity so that, following its recent recovery, productivity will grow until the end of the century. Quite probably the Board is correct in assuming some increase in productivity as conventionally measured by dividing output per time-period by the number of miners or man-shifts. If new capacity is relatively capital-intensive, however, the conventional measures will become misleading since the real input of productive factors will be more than appears from the coal industry's traditional productivity statistics.

Earnings, productivity and bargaining power

Whether or not there will in practice be a significant change in the proportions of labour and capital employed in coal mining is an interesting question which we can address indirectly by considering the likely trend of earnings in relation to labour productivity (as traditionally measured). We have already seen that in the 1970s NCB mining operations became *more,* not *less,* labour-intensive as labour costs rose relatively fast (Table XII). For understandable reasons, the Coal Board and the Department of Energy say little in public about miners' future earnings. Consequently, there is some tendency for the subject to be played down in public discussion and in official documents about energy policy, where simplistic statements about future labour costs abound. In the 1978 Green Paper, for example, the Department of Energy, in discussing the Selby coalfield and other new capacity, stated that

'The new capacity will be capital-intensive and the proportion of wages costs to total costs, currently 50 per cent, should be substantially reduced.'[1]

Whether or not such statements come true depends not only on investment in coal but on the actions of the mineworkers and their union.

It is clear that the bargaining power of the National Union of Mineworkers will be very considerable in the foreseeable future. The mining labour force is still very large at nearly a quarter of a million and the heavy dependence on coal of the

[1] Cmnd. 7101, para. 6.5.

electricity supply industry (over three-quarters of its fuel is coal) magnifies the impact of industrial disputes in the coal industry. During a period in which the coal industry is trying (whether successfully or not) to expand, the miners will be in a strong position to achieve much higher earnings. We are not, of course, concerned in this *Paper* with what miners' earnings *ought* to be but with what *is likely* to happen to them. Workers in the industry clearly feel that, after their long period of wage restraint during the years of coal decline in the 1960s, they should receive big increases during the anticipated expansion of coal; rising oil prices will appear to give them plenty of 'head room'.

An indication of the miners' bargaining strength is given by experience in the early 1970s when there were big rises in world crude oil prices. Between 1973-74 and 1976-77 labour costs per tonne in NCB mines almost doubled (Table IV) and the price of coal used by industry also doubled (Table V), so that there was a significant erosion of the initial competitive gain to the coal industry from the oil price increases. During the years of rigid incomes policies the bargaining power of the miners was temporarily concealed, but between 1978 and 1980 substantial increases in earnings were obtained. In December 1979 the miners gained a pay increase of about 20 per cent for the 10 months ending December 1980, and at the end of 1980 they settled for a rise in earnings of approximately 13 per cent for the 10 months to October 1981.[1] On both occasions, industrial action seemed possible but was averted by coalfield ballots. In 1981, when the miners' pay claim will be from its traditional date of November, any threat of industrial action will, other things equal, be more serious since it would start at the onset of winter.

The pay increases of the last two years, together with rising non-labour costs, have resulted in coal prices rising substantially. In March 1980 the average price of NCB coal went up by about 19 per cent;[2] in November 1980 domestic coal prices increased by 10 per cent;[3] and in January 1981 industrial coal prices rose 11 per cent and domestic prices 8 per cent.[4] The

[1] 'Miners accept 9·8% and strengthen government position', *Financial Times*, 2 December 1980. The '9·8%' refers to the increase in basic rates, not earnings.

[2] 'Industry coal price up 20%', *Financial Times*, 15 February 1980.

[3] 'Coal prices to rise 10% next month', *Financial Times*, 3 October 1980.

[4] Department of Energy, *Energy Management*, January 1981.

Coal Board has an agreement with the CEGB to supply 75 million tonnes a year of power-station coal during the period 1978-83 provided its prices do not rise faster than retail prices generally. Perhaps because of that agreement, increases in coal prices to the CEGB have been kept just about in step with the rate of retail price inflation (on a somewhat charitable CEGB interpretation of what that rate is). Clearly, cost trends in the coal industry are making it difficult for the NCB to keep within the agreement, and the CEGB, which is under financial pressure from the Government, must have an incentive to increase its imports even more than it did in 1980 (Section VI).

The long-term outlook for coal costs and prices

More important than the immediate prospect is the long-term outlook for costs and prices. It seems to us difficult to escape the conclusion that in the future, as in the past, there will be a strong link between oil prices, miners' earnings, other costs of mining and the price of British coal.

First, it is generally believed that oil prices will rise substantially. Although the only correct statement we can make about past forecasts of changes in oil prices is that they have almost invariably been wrong, there are on this occasion some reasons of substance why prices are likely to increase: principally, conflicts and tensions in the Middle East may lead to periodic reductions in oil supplies; uncertainty about the production plans of major OPEC producers may drive prices up; and oil reserves recoverable at or near present costs will probably become scarcer.

But, in the absence of freedom to import coal, the British energy consumer will be subject to the monopoly power of two groups, not one. The activities of the oil producers receive considerable attention in the media where they are frequently portrayed as grasping foreigners. However, the British coal industry also enjoys considerable monopoly power in circumstances of rising oil prices.[1]

If Britain had a number of competing coal suppliers (domestic or overseas), competitive pressures would tend to keep down coal prices as oil prices rose so that coal's market share would expand. Given the semi-monopolistic position of the industry,

[1] Colin Robinson, *The Energy 'Crisis' and British Coal, op. cit.,* p. 44.

[56]

however, and the very strong bargaining power of the NUM, it would be surprising if workers and managers in the coal industry did not succeed in capturing much of the market surplus (or rent) resulting from higher oil prices. Non-labour costs also seem likely to rise substantially because the NCB will have a reduced incentive to be cost-conscious as competitive pressures are lifted by the rise in price of other fuels.[1] As costs increase, the NCB should find it comparatively easy to pass them on to consumers in higher prices. Thus there is a strong probability that oil and coal prices will be closely related.

How precisely the link will work is open to question. The NUM, for example, has a choice between raising wages sharply so that, with other costs also increasing, coal prices stay at about parity with oil (after allowing for the disadvantages in use of coal). Or it might decide to raise earnings less and, hoping that other NCB costs will be held down, take the gains in terms of employment. What proportion of the rent will be captured by the coal industry is a matter of opinion. Our view is that it would be consistent with recent experience to assume that the NUM will opt primarily for higher earnings, that other coal industry costs will also increase considerably, and that, in consequence, over a long period (though not necessarily in the short term) the price of British coal will tend to approach the ceiling set by oil prices. Thus the British coal industry would capture most of the rent arising from the increase in prices of other fuels, just as in the last few years British Gas has taken the rent from North Sea gas.[2] Our view is based on the assumption that, for one reason or another, coal imports will be small. If coal imports become more significant there will, as we explain in Section VI, be a more competitive market for coal with imports effectively setting the energy price level.

---

[1] As an example of how 'organisational slack' can develop in an industry which has relatively little competition, one has only to consider the behaviour of the British Gas Corporation in the period since it was given access to low-price natural gas from the North Sea. Colin Robinson and Jon Morgan, *North Sea Oil in the Future: Economic Analysis and Government Policy*, Macmillan for the Trade Policy Research Centre, and Robinson, 'A Review of North Sea Oil Policy', *Zeitschrift fur Energie Wirtschaft*, 4/1978.

[2] Robinson and Morgan, *ibid.*

A similar view of the likely competitive position of coal, in the absence of substantial imports, is taken in the Electricity Council's 1979 *Medium Term Development Plan*:

> '. . . it would not be reasonable to expect that the rate of increase in UK coal prices will be significantly less than that made possible by increases in world oil prices'.[1]

It is only fair to add that at the Belvoir Inquiry the Central Electricity Generating Board differed from the Electricity Council's view. The reasons were not entirely clear. The CEGB's witness said on 14 January 1980 that the electricity supply industry had changed its assumption about the rise in real oil prices by the end of the century: instead of assuming they would double (as in the Development Plan) they were assuming a 2·7 times increase. But two weeks *later,* on 30 January 1980, a CEGB memorandum to the House of Commons Select Committee on Energy stated that one of its 'key cost assessments' concerned oil prices which

> 'are widely anticipated to rise sharply and CEGB has assumed broadly a *doubling* of prices by the year 2000' (our italics).

Quite apart from this unfortunate inconsistency, we are quite unable to understand why the basic economic mechanism which apparently connects oil and coal prices should be changed because of a different assumption about *one* of them.

The Department of Energy's views about the future relationship between coal and oil prices are also interesting, if only because they reveal assumptions which seem even more curious than those used in their forecasts of total energy demand, which appear to us too high (Section III). *Energy Projections 1979* has very little to say about relative prices. There is a single assumption about world oil prices: they will rise about $2\frac{1}{2}$ times in real terms by the end of the century (apparently from 1977). The only information given about coal is that

> '. . . costs were arrived at by making projections of productivity and labour and other costs in the UK coal industry'

and that

> '. . . miners' wages will be maintained at a level significantly above

[1] Electricity Council, *Medium-Term Development Plan 1979-86,* 1979, para. 98.

[58]

the national average and . . . other operating costs will rise in real terms to the end of the century'.[1]

According to the Department, no details could be disclosed to the Inquiry because the information is too sensitive. Such an attitude may be understandable, but the Department was hardly able to make the same excuse for not explaining its views on *relative* prices.

The Department's relative price assumptions, when revealed in a letter to the Inquiry Secretariat, contained a number of surprises. First, it uses a concept, difficult to comprehend and never properly explained to the Inquiry, of the *cost* of coal relative to the *price* of oil. The only conclusion we could reach was that the 'cost' of coal excluded some or all of the rent which the coal industry might be expected to extract from the market as a consequence of rising oil prices. Whatever this unusual concept of 'cost' is supposed to mean, the Department of Energy appears in *Energy Projections 1979* to have used the coal cost/oil price ratio as though it were a relative *price*.

Once this procedure is revealed and the Department's assumptions are known, one of the main reasons why *Energy Projections 1979* estimates an increase in coal demand becomes clear. The projections suggest that coal consumption in the year 2000 will be in the range of 128 to 165 million tonnes—less than the Coal Board's 170 million tonnes but above the 1977 figure of about 123 million tonnes which is the base for the Department's forecasts. To obtain this result the Energy Department uses cost/price assumptions which seem extremely favourable towards coal:

COAL COST/OIL PRICE RATIOS ASSUMED BY THE DEPARTMENT OF ENERGY

1985	0.6
1990	0.5
1995	0.45
2000	0.4

The concept of coal cost is unclear; nor do we know how the price of oil has been measured. If, however, the ratios are compared with those in Table V, on the assumption that they

[1] *Energy Projections 1979*, para. 6 and 12.

relate to industrial prices, they are well below the experience of recent years except for 1974, which was just before a catching-up period for coal prices, and early 1980, which was also probably just prior to a rise in the relative price of coal.

No explanation of these apparently extreme assumptions has been provided by the Department of Energy. It is obvious that a forecast based on assuming a marked and uninterrupted downward trend in the relative price of coal for a period of about 20 years, and assuming also an expanding total fuel market, will show a significant rise in coal consumption. The question is: can the assumptions be justified?

It seems to us quite unrealistic to assume that the 'cost' of coal (which seems to have been used as a price) will sink steadily until, by the year 2000, it is only 40 per cent of the price of oil. The implication of the Department's assumption is that neither the NUM nor the NCB will take advantage of the rise in price of other fuels to extract the rent which could accrue to the coal industry. Such behaviour is conceivable but unlikely. Indeed, the Department's relative price assumption seems so at variance with past experience that those who put it forward as one of the principal bases for projections of British fuel demand have a responsibility to explain it. As part of a sensitivity test to determine *what would happen if* coal prices were to fall lower and lower relative to oil prices it would be interesting. But the Department has carried out no such tests. It had to confess to the Belvoir Inquiry that

'The cases in the Projections were constructed on the oil price and coal cost assumptions specified in the document and in evidence. No alternative coal cost/oil price ratios were applied to them as sensitivities.'[1]

The Department has to be given full marks for honesty but very little for forecasting technique: most forecasters regard sensitivity analysis—to test the effects of varying key assumptions—as an indispensable part of forecasting. It is also usual for professional forecasters to state their assumptions clearly, rather than revealing them subsequently and rather unwillingly to outside inquirers. Given the uncertainties of the energy market and the importance of relative prices, testing the effects of changing clearly-specified price assumptions is one of the most essential features of any energy forecast.

[1] Letter to Belvoir Inquiry Secretariat dated 18 January 1980, p. 4.

Conclusion: coal projections unrealistic

We can only conclude that the coal estimates in *Energy Projections 1979* constitute neither a realistic nor an objective view of the future of indigenous coal (or of any other fuel) in the British market. There are reasons to believe that the forecasts of the total market for primary fuel are on the high side (Section III). Moreover, within that total market, the share of coal is probably substantially over-stated because of the extreme assumption made by the Department of Energy of a falling relative 'cost' of coal. We set out in Section V some of our own estimates of future coal consumption, based on various relative price assumptions, and make comparisons with the projections of the Department of Energy and the NCB.

V. PROSPECTS FOR BRITISH COAL:
WHAT WILL BE ITS MARKETS?

Department of Energy and NCB estimates

We begin our assessment of the future markets for coal by examining the estimates made by the NCB and the Department of Energy (Table XIV). We have explained in general terms (Sections III and IV) why the Department's estimates appear to be too optimistic. Their most striking feature is the very large projected expansion in the industrial market to between five and six times its present size by the year 2000. Industrial coal sales have more than halved since 1970 (Table III); thus a remarkable turn-round in this market is forecast. In other markets, the Department projects a decline in coal sales to power stations and an increase to coke ovens. The reason the upper end of the Department's range for other consumers ('Domestic/Commercial') shows an increase compared with the present is that it includes a wide range (1-15 million tonnes) of possible coal use for substitute natural gas (SNG) manufacture.

Both the upper and lower ends of the Department's ranges for total coal consumption in the year 2000 are below the corresponding NCB estimates (shown in the second and third columns of Table XIV). The status of the NCB's figures is unclear. The ranges given in *Coal for the Future* (column 2) were published early in 1977, but the estimates given by Sir Derek Ezra in 1979 to the Commission on Energy and the Environment (column 3) are around the mid-points of the *Coal for the Future* ranges. It might therefore be concluded that the Board has not changed its views in the last three or four years on future coal consumption. That conclusion would be consistent with some of the NCB's statements at the Belvoir Inquiry (Section I) which suggested that it is still aiming to produce 170 million tonnes or thereabouts at the end of the century.

However, from the very limited amount of information given in *Coal for the Future* about how the estimates of coal demand by market were arrived at, it seems scarcely conceivable that the NCB can continue to use them as realistic projections. All

[62]

TABLE XIV

ESTIMATES OF POSSIBLE MARKETS
FOR COAL IN 2000

	1979 Actual	National Coal Board (1977)*	National Coal Board (1979)†	Department of Energy (1979)‡
			million tonnes	
Power stations	90	75- 95	90	66- 78
Coke ovens	12	20- 25	20	16- 19
Industry	8	30- 50	40	39- 45
Domestic/Commercial (including substitute natural gas)	14	10- 30	20	7- 23
	124	135-200	170	128-165

*Coal for the Future, para. 48.

†Statement by Sir Derek Ezra on 16 January 1979 to *Commission on Energy and the Environment*.

‡*Energy Projections 1979* and explanatory letter from Department of Energy to Vale of Belvoir Inquiry Secretariat, 18 January 1980.

those estimates were based on a forecast of total consumption of primary fuel in the year 2000 (500 to 650 mtce or more) which virtually no-one now believes (Section III). Moreover, the assumptions used to estimate individual markets are obviously dated. For example, *Coal for the Future* assumes that total fuel requirements in the power generation market in 2000 will be 160 to 200 mtce or more. Estimates made by the CEGB in February 1980 suggest, after allowance for the other electricity boards, that the figure will be 130 to 140 mtce.[1] Similarly, *Coal for the Future's* estimates of coal demand for coke ovens assume steel output of no less than 35 to 42 million tonnes in 2000 whereas Department of Industry estimates now put the range at 21 to 25 million tonnes (Section III).

The Board's forecasts for the industrial market assume a very big expansion of coal sales (comparable in scale to the Energy

[1] Letter to Belvoir Inquiry Secretariat dated 28 February 1980.

Department's estimate), and seem to assume an even faster growth in British manufacturing production than the Department's figure criticised in Section III. The Coal Board's forecast of manufacturing output is not stated, but it must be extraordinarily high since it assumes a growth in the total industrial market for fossil fuels of 40 to 80 per cent, as compared with the Energy Department's 15 to 30 per cent.

At the time of the Belvoir Inquiry, it seemed to us that neither the Department of Energy's estimates of coal consumption nor those of the NCB were of any practical value as a guide to the future. The Department's projections contain so many inconsistencies and unrealistic assumptions which bias them upwards that they can be regarded only as an extreme view of coal's future. The Coal Board was unwilling to present any up-to-date demand estimates to the Inquiry, and those it has published in the past seem based on assumptions which must make them misleading. At the Inquiry the Board preferred to rely on Department of Energy assumptions and forecasts (page 32). Consequently we prepared our own forecasts of coal consumption. Like all forecasts, they are fallible and subject to wide margins of error. But their purpose is not to show precisely what will happen in the future—which is impossible —but to demonstrate what broad trends are likely, what uncertainties exist, and how sensitive the conclusions are to changes in key factors such as relative prices and the rates of growth of macro-economic variables.

The power generation market

The power generation market, which accounts for over 70 per cent of NCB sales and has been its only growing market in recent years, is clearly crucial to the future of British coal. The NCB expects its sales to power stations in 2000 to be in the range of 75 to 95 million tonnes, or about the same order of magnitude as in 1980, although it gives no detailed explanation of how the range is calculated (Table XIV). The Department of Energy projects a range of 66 to 78 million tonnes—13 to 27 per cent less than in 1980—but uses the assumption we have criticised of a big fall in the relative price (or 'cost', to use the Department's terminology) of coal and does not examine the effects of varying its assumptions about relative prices.

Our forecasts of coal consumption by power stations are explained in the Appendix (pp. 85-101) which shows how, from specified assumptions about electricity growth, power station construction and closure plans and the relative prices of coal and oil, estimates can be derived of the quantity of coal the electricity supply industry will burn. The range of consumption estimates produced by our scenario/sensitivity analysis approach is inevitably wide; indeed, one of its objectives is to show areas of uncertainty rather than to calculate spuriously precise 'point forecasts' or narrow ranges. The broad conclusion of the Appendix is that sales to power stations are likely to decline significantly between now and the end of the century. The Coal Board's *minimum* estimate of 75 million tonnes in 2000 is well above the *maximum* calculated by us. What exactly will happen to sales for electricity generation is partly a matter of judgement but, in the light of the likely competitive position of coal (without the threat of significant imports), we would deduce from the Appendix a probable range for the end of the century of 40 to 60 million tonnes, or about 30 to 50 million tonnes less than in 1979. It is possible, using our methodology but the Department of Energy's assumptions, to obtain a range similar to the Department's (66 to 78 million tonnes). But those assumptions are that the price of coal (or the 'cost' of coal) will fall to about 40 per cent of that of oil (an entirely implausible assumption for long-term forecasting if the coal is primarily British), and that electricity sales will increase at about 2 per cent a year to the end of the century. The latter assumption is in line with the CEGB's medium-term projections at the time the Belvoir Inquiry started, but those projections are now for only 1 per cent growth (Appendix).

Before and during the Belvoir Inquiry the CEGB produced several estimates, differing widely from each other, of the probable coal consumption of power stations in 2000. Its 1978 Corporate Plan has a figure of 30 to 47 million tonnes; at the Inquiry it suggested 72 to 106 million tonnes; and, after the very large downward revision in February 1980 of its forecasts of electricity demand, it estimated 45-79 million tonnes. For comparison with our figures of 40 to 60 million tonnes, the CEGB estimate[1] of February 1980 is equivalent to about 50-87 million tonnes for Great Britain as a whole (the CEGB figures are for England and Wales only). Thus it is fairly similar to ours at the bottom of the range but much higher at

[1] The Appendix (page 89-91) explains that there may be a further reduction in the CEGB's forecasts of electricity consumption.

[65]

the top. The latest CEGB estimates are based on a 1 per cent a year growth of electricity demand and a coal/oil price ratio in 2000 of about 0·6. The top end of the range for coal consumption appears to assume not only this price advantage for coal but a nuclear programme similar to our 'Nuclear Delays' scenario (Appendix, p. 95).

Other markets for coal

Apart from power generation, coke ovens and sales to industry will probably be the only two substantial markets left to the coal industry at the end of the century, although the Department of Energy also sees the prospect of a new market for coal in SNG manufacture. Of the other markets where sales are now fairly large, even the NCB and the Department agree that 'Domestic and Commercial' will be much reduced by the end of the century. The estimates we made for the Belvoir Inquiry put coal sales to these two markets combined at 5 to 7 million tonnes in the year 2000, which is only fractionally below the Department's estimate of 6 to 8 million tonnes (excluding the SNG market). Export markets for British coal seem unlikely to revive, despite the prospect of a considerable increase in world coal trade. British exports have been between 1 and 4 million tonnes annually in recent years and the NCB has little chance of exporting significantly more in competition with the cheaper products of the United States, Australia and South Africa. The rise in exports in 1980 seems to have resulted from loss-making sales which the NCB will not be able to continue indefinitely.[1]

Future coal sales to coke ovens will depend largely on the fortunes of the British steel industry. Our estimates of coal sales to this market are lower than those prepared by the NCB and the Department because we take a less optimistic view of the steel industry's future. The NCB's estimates (Table XIV) use the hopelessly outdated assumption that British crude steel output will almost double to between 35 and 42 million tonnes by the year 2000. The Department of Energy is a little more realistic, using a figure for the year 2000 of 26 to 30 million

[1] *Steam Coal, op. cit.*, states (p. 111): 'The high delivered cost of British coal is expected to preclude exports'. According to 'NCB plans port expansion to treble its coal exports', *Financial Times*, 10 March 1981, the NCB hopes to raise exports substantially in the next few years even though they are not profitable.

tonnes in *Energy Projections 1979*. In our own estimates we took a figure of between 20 and 22 million tonnes, which is slightly below the Department of Industry's latest assessment of 21 to 25 millions (Section III). Our estimate of coal sales to coke ovens in 2000 is between 13 and 15 million tonnes compared with the NCB's 20 to 25 millions and the Department's 16 to 19 millions.

Bigger differences emerge in our views of the future industrial market for coal. The NCB and the Department both expect a very large increase in sales to industry. The only attempt to explain the NCB's estimates is in *Coal for the Future* which however is insufficiently explicit for anyone to be able to understand how they were made. The Department of Energy is a little more specific. In *Energy Projections 1979* it assumes the remarkable recovery in British manufacturing on which we commented in Section III and consequently predicts a considerable growth in the industrial fuel market—from 16·4 billion therms in its base year of 1977 to between 21 and 24 billion in the year 2000. In addition, it makes an assumption about natural gas sales to industry which we regard as rather extreme: in 2000 they are projected at only half their 1977 total. This low estimate is consistent with the Department's generally pessimistic line in *Energy Projections 1979* about future natural gas supplies. The arithmetical approach of the Department then indicates that an expansion in the total market, accompanied by falling gas supplies, leaves a 'gap' for other fuels to fill. Inevitably, given the relative price assumptions for coal and oil the Department uses (Section IV) and its optimism about the spread of new coal combustion technologies, it reaches the conclusion that coal consumption by industry will increase rapidly to fill this supposed gap.

The estimates we made for the Belvoir Inquiry suggest a more modest increase in industrial coal sales. We assumed (Section III) that, with real GDP growing at between 2 and 2½ per cent a year, UK manufacturing production will increase by 1 to 1½ per cent annually to the year 2000, compared with the Department's 2 to 2·7 per cent.[1] Thus the total industrial fuel market will increase relatively slowly on our assumptions (Table IX). Within that total, coal also has a smaller share than in the Department's estimates because we do not use its

---

[1] If anything, our estimates of manufacturing output may be over-optimistic since we are less confident now than when the Belvoir Inquiry started that 2 to 2½ per cent annual growth in real GDP will be achieved.

extreme assumption about a fall in the relative price of British coal but take the same range of relative coal/oil prices as is shown in the Appendix.

Obviously, there is a large element of speculation in estimates of future industrial coal sales. We simply attempted a systematic analysis in which we modelled the effects on the industrial coal market of changes in manufacturing output, relative prices and the efficiency of coal use; the last variable is included because new coal combustion technologies (such as fluidised bed combustion) will raise the efficiency of coal utilisation. On a range of assumptions which seemed to us reasonable, the highest figure for coal sales to industry we could estimate for the year 2000 was about 25 million tonnes, compared with the *low* ends of the NCB and the Department ranges of 30 and 39 millions respectively.

NCB and Department of Energy's 'extreme optimism'

Having attempted some estimates of our own, our strong suspicion is that the NCB and the Department of Energy are using estimates for industrial coal sales which, like those they make for the power generation market, are at the optimistic extreme of what can be expected. Only if manufacturing expands considerably, if coal prices are maintained at less than half those of oil, and if new coal combustion technologies spread fast are their estimates likely to be approached. Probably the NCB is caught up in enthusiasm for new combustion technology and expects it to spread far more rapidly than is realistic. It is likely that neither the Board nor the Department has given enough weight to the serious practical obstacles in the way of such a sharp revival of industrial coal sales as they envisage. Many firms, for example, no longer have the storage, handling and transport facilities for coal which they had before oil took over the market; it will be time-consuming to replace them. Moreover, the cost of conversion to coal seems to be such that some firms consider it uneconomic at the relative coal/oil prices they anticipate.[1]

Finally, there is the market which both the NCB and the Department of Energy claim will develop for the manufacture of SNG from coal. *Energy Projections 1979* states that

[1] 'NCB launches campaign for plant conversions', *Financial Times*, 25 September 1980; 'British Sugar set to reject EEC's £0·75m energy offer', 31 October 1980· and 'Coal Board urges grants for switch to coal-firing', 8 January 1981.

'. . . before the end of the century there could also be a requirement for manufacturing substitute natural gas from coal',[1]

but it does not explain what the *size* of such a market might be.
It emerged from cross-examination at the Belvoir Inquiry that
Energy Projections 1979 assumes sales of coal for SNG manufacture in the year 2000 of between 1 and 15 million tonnes. The
Inquiry also discovered that the 1978 Green Paper had included
a figure for the year 2000 of about 29 million tonnes for SNG.
Although the Department has drastically reduced its estimate
of end-century coal consumption in this market, the upper
end of its range (15 million tonnes) still seems extremely high
to us. Its high estimate is probably a reflection of its pessimism
about North Sea natural gas supplies: it assumes they will
dwindle quite sharply in the 1990s.[2] Of course, no-one can be
sure what will happen to gas supplies in the next 20 years. In
our view, with rising oil and gas prices stimulating exploration
activity and the prospect of gathering gas from relatively small
pockets associated with Northern Basin oil fields, there is no
reason to expect the decline in the 1990s the Department
assumes—unless government pricing and tax policies take
away the incentive to expand supplies which rising prices
would otherwise provide. In our own estimates we included a
token 2 to 3 million tonnes of coal consumption for SNG in
the year 2000 simply to indicate that a new small market for
coal might by then have appeared.

Coal markets in the year 2000 and beyond

Our estimates of coal consumption by market in the year 2000
are summarised in Table XV and compared with the 1979
estimates of the NCB and Department of Energy. It is assumed
that coal imports will be limited to the relatively low levels of
recent years.[3] The estimates are the same as those we made for
the Belvoir Inquiry, except for the power generation market
which we then suggested would be 45 to 65 million tonnes in
2000.[4] The implication of our calculations is that, far from
the considerable expansion anticipated by the NCB and the

---

[1] *Energy Projections 1979*, para. 13.

[2] *Energy Projections 1979*, para. 15.

[3] The effect of unconstrained imports is discussed in Section VI.

[4] Our reasons for changing our estimates of the power generation market are
given in the Appendix.

[69]

TABLE XV

PRESENT AND FUTURE COAL MARKETS

	1980 actual	NCB estimates*	Year 2000 Estimate Department of Energy estimates†	Our estimates‡
			million tonnes	
Power stations	90	90	66- 78	40- 60
Coke ovens	12	20	16- 19	13- 15
Industry	8	40	39- 45	15- 25
Domestic/ Commercial (including SNG)	14	20	7- 23	7- 10
	124	170	128-165	75-110

*Table XIII, column 3.
†Table XIV, column 4.
‡Based on assumptions explained in Section V.

more modest increase the Department expects, a decline in coal consumption is likely during the next 20 years. The best position we can foresee is a fall in coal sales of about 11 per cent.

What is likely to happen to coal consumption beyond 2000 is of course extremely uncertain. But it is of some consequence, bearing in mind that investment projects such as the proposed Belvoir mines involve sinking pits which may produce coal for 50 to 100 years.

The Department's view of the future of coal beyond 2000 appears to be that there is a prospect of considerable expansion. At the Belvoir Inquiry the Department's witness said that the annual demand in the next century might be of the order of 200 million tonnes. At the same time he claimed that sales for power generation would be 'negligible', implying an enormous rise in sales to other markets since the Department's projections show the power generation market in 2000 at 66 to 78 million tonnes, or about half total coal sales (Table XV). No indication was given of how these estimates were made, nor of the economic growth and relative price assumptions used. They appear to us to be extraordinarily high.

A point of agreement among most witnesses at the Belvoir Inquiry who were concerned with demand (including those from the CEGB and the Department) was that the power generation market for coal would reach its peak around 1990 and decline quite rapidly into the early part of the next century. It is very difficult to see how the Coal Board will be able to find new or revived markets of sufficient size to compensate for the contraction of the electricity generation market—let alone permit the significant expansion the Department suggests—unless British coal becomes more competitive than we have argued is likely. Consequently, we suggested at the Inquiry that coal demand would probably still be falling in the early part of the 21st century.

VI. THE POLICY ISSUES

A general view of the coal estimates

We have argued that the published expansion plans of the National Coal Board are based on out-of-date assumptions which make them look extraordinarily optimistic. Estimates of future coal consumption prepared by the Board's sponsoring Ministry, the Department of Energy, also appear to rest on unrealistic assumptions—particularly about the growth of British industry and the competitive position of British coal— which give them a significant upward bias. The estimates we have made do not pretend to give other than a very broad view of the future of coal; no-one should expect 'forecasts' to the end of the century to do more. Nevertheless, they set out what seem to us plausible ranges of consumption, market by market, which, despite including a substantial increase in the tonnage sold to industry, point towards some decline in total coal consumption. The scale of the decline to the year 2000 is a matter of conjecture; our estimates suggest a range of 11 to 40 per cent as compared with 1980. We see no evidence that an upturn in the industry is likely in the early years of the next century.

A number of policy issues arising out of our analysis and the Belvoir Inquiry are discussed below. Some of them concern the suitability of the Public Inquiry procedure for dealing with major energy investment projects such as the proposal to mine coal in North East Leicestershire; the others relate to the place of coal in the British energy market.

The Public Inquiry procedure

Anyone who observed the protracted Public Inquiry into coal mining in North East Leicestershire cannot but have been impressed by the cumbersome nature of the procedure and the infinite variety of the matters discussed. In this *Paper* we have been concerned with only one aspect of the Inquiry. However, of the total of well over 100 witnesses who appeared during the Inquiry, which began on 30 October 1979 and ended on

[72]

22 April 1980, only five were concerned directly with the so-called 'need' for Belvoir coal, although three more presented evidence peripheral to the 'need' case. The rest of the witnesses dealt either with matters of detail related to the NCB's proposed methods of extracting and transporting the coal, or with the scale of operations (such as the number of miners and the amount of investment required), or with major environmental aspects (for instance, where the large amount of waste would be tipped and how much subsidence there might be), or with the effects on particular individuals of the Board's plans. The consensus among those who attended the proceedings was that they were tedious and that the material produced was so diverse as to be indigestible, presenting in the end a confused mass of evidence from which it must have been extremely difficult to conclude whether or not the three proposed pits should be allowed.[1]

It seems to us unreasonable and unnecessary to place on the Inspector at a Public Inquiry the burden of sifting and weighing such a diversity of views on such a variety of subjects. The Belvoir Inquiry was fortunate in having a presiding Inspector who could appreciate arguments presented on a wide range of subjects: in other cases Inquiries may be less fortunate.

An alternative procedure which has some merits would be to hold such Inquiries in two stages. The first stage would discuss the National Coal Board's expansion plans, examining in detail the underlying assumptions and allowing both government departments and individuals who had studied the prospects for coal to comment. The intention would be to bring the debate about coal into the open rather than allowing it to be settled in private between the NCB and the Department of Energy. At such a first stage it would be much more difficult for the Coal Board to obfuscate its ideas about future output and consumption than in the midst of the bewildering variety of matters discussed at the Belvoir Inquiry. To the end of that Inquiry we could not determine whether the Board still believed in the estimate it had suggested in 1977 of coal output and sales of 170 million tonnes by the end of the century. The report of the first-stage Inquiry would presumably include a review of the key variables likely to affect the future of British coal and some broad conclusions about that future which would

[1] 'Half Time in the Vale of Belvoir', *New Scientist*, 21 February 1980.

clear the ground for local Inquiries into NCB applications to mine particular areas, such as the Vale of Belvoir or Park[1] in Staffordshire. The local Inquiry would, in effect, be a cost/benefit analysis of whether a particular area was the most appropriate site for the next major mining development, taking into account both private and social costs. A procedure, first created by the Town and Country Planning Act of 1968, already exists for a two-stage Inquiry, known as a **Planning Inquiry Commission (PIC)**. The Leicestershire County Council unsuccessfully proposed such a procedure for the Belvoir Inquiry. To date, no PIC has ever been appointed.

As presently formulated, a PIC would seem not to be an ideal forum for investigating the many-sided and wide-ranging implications of a project like Belvoir. This is not because the procedure has two stages, but because the statute requires the second local Inquiry to be carried out by an Inspector who *must* have been one of the Commissioners for the first stage. He will, therefore, have formed views on policy and principle which may not be known to those who participate in the second stage; consequently the participants may lack confidence in the Inspector's decisions. Nor will other institutions, such as government or Parliament, have had the opportunity of considering or taking a decision about these policy matters. Within the statutory PIC procedure there is no way round this problem. Moreover, for issues of national concern, the PIC has another disadvantage. As with ordinary local planning inquiries, it is 'triggered' by an application for planning permission; it cannot be set up until the project is far enough advanced for the applicant to have taken that step, which may well be a good deal later than others would consider the best time.

Some other procedure must therefore be found for investigating such large projects with national and international implications and involving important environmental as well as economic issues. This conclusion is not new and is apparently shared by the major political parties.[2]

---

[1] Press reports indicate that the NCB may not proceed with Park because of the high chlorine content of the coal (*Financial Times*, 29 December 1980).

[2] For instance, the statement by the (then) Secretary of State for the Environment on 13 September 1978: Department of the Environment Press Notice No. 488; and the Conservative Party's 1979 Election Manifesto.

More important than the procedures used at Public Inquiries are the structure of the British energy market and the conduct and performance of the NCB in that market. Most people who have practical experience of forecasting will have reservations that Public Inquiries—even if held in stages to separate the issues in some logical fashion—may become a contest between opposing forecasts. Although we have no hesitation in arguing that the evidence points to excessive optimism in the coal forecasts of the NCB and the Department of Energy, we have little faith in any attempt to state exactly what the future of coal will be. It seems to us more sensible to establish a *system* which will produce a closer approximation to the 'correct' amounts of coal investment and output than to try to guess within the present system what that output will be and then impose the investment. In general, it seems to us unlikely that reasonable decisions about coal investment will be taken so long as the British coal industry retains the amount of monopoly power it enjoys at present. The industry needs to be placed in a more competitive environment so that it can be more readily judged whether investing in new pits, in the Vale of Belvoir or elsewhere, is a wise use of productive factors or a misallocation of resources.

One serious difficulty at the Belvoir Inquiry was the uncertainty over how much competition the NCB would be permitted to face in future. A major element in the case we presented at the Inquiry (Section IV) was that the semi-monopoly power of the British coal industry would result in a link between oil and coal prices such that, assuming oil prices continued to rise, the price of indigenous coal would in the long run increase roughly in step. The link between world oil and British coal prices could, however, be weakened by a reduction of the monopoly power of the British coal industry. If, for example, it was believed that coal could be imported without restriction into the UK and if those imports would be cheaper than oil or British coal, the NCB's ability to raise prices would be severely constrained. There would be significantly more competition in the UK energy market; and energy prices would probably be set by imported coal, as they were set by oil in the 1960s. In such circumstances coal consumption in the UK might well be significantly higher than our estimates, which assume that imports will be constrained (Table XV). Because of the

[75]

potential importance of imported coal we consider it in more detail below.

Recent trends in coal imports

A recent trend in the British coal market is for imports to increase (Section I). Annual imports were in the range of 2 to 3 million tonnes between 1976 and 1978, but increased to about 4½ million tonnes in 1979. There was a very large further rise in 1980 to over 7 million tonnes.[1] Both the British Steel Corporation and the CEGB have been importing more.

According to press reports, British Steel at one time threatened to buy all its coking coal from abroad in 1981-82 unless the NCB agreed to reduce its price to be competitive with imports.[2] The dispute was settled only when the NCB undertook to cut its prices to BSC. Even so, BSC will buy only about 4 million tonnes of coking coal from the NCB in 1981-82, which is approximately the same as in 1980-81 but much less than the 7 millions of 1979-80 and 8½ millions of 1978-79. It appears that, in the summer of 1980, foreign coking coal could be bought at about £10 per tonne less than the NCB product.

The electricity supply industry plans to import about 4 million tonnes in 1980-81 and is clearly anxious to have the capacity to import more. The Electricity Council's *Medium Term Development Plan 1979-86* (para. 33) states that '. . . the industry should participate in the growing international trade in coal'. Practical steps are already in train to allow the CEGB to import more. It was disclosed at the Belvoir Inquiry and subsequently reported in the press[3] that the CEGB, which at present has the capacity to import about 5 million tonnes of coal a year, is seeking one or more ports where facilities to handle annual imports of about 10 million tonnes can be constructed.

It is obvious that, for both the BSC and the CEGB, there are great advantages in coal imports. For example, they provide a flexible means of coping with any short-lived supply problems of British coal. Moreover, long-term import contracts might well be advantageous in terms of price. They would improve the security of supplies by diversifying supply sources,

[1] Department of Energy, *Energy Trends*, January 1981.

[2] '£40m coal subsidy helps BSC buy British', *Financial Times*, 6 September 1980.

[3] 'Ports plan may lift coal import capacity to 15m tonnes', *Financial Times*, 18 February 1980.

and would provide competition for the NCB, so constraining its ability to raise prices.

Similar advantages would, in our view, accrue to the British economy as a whole if the NCB were placed in a more competitive environment in which low-cost imports of coal imposed strong pressures to operate efficiently and prevented the Board from simply raising its prices *pari passu* with increasing oil prices. In the 1960s the coal industry was subjected to competitive pressure from the falling real price of products refined from imported crude oil, and was thus unable to exploit the monopoly powers granted it by the state. Now that the ceiling has been lifted, another constraint is necessary; in its absence the British energy consumer will be faced by two semi-monopolistic groupings in the forms of OPEC and the British coal industry (Section IV).[1]

The world coal market and coal prices

We quoted in Section II (page 38) an extract from a statement by the NCB that world coal prices would be 'greatly affected by the rising world price of oil'. At the Belvoir Inquiry the Board maintained that the price of imported coal was at present artificially low and would increase as oil prices rose. These views appear to be inconsistent with others expressed by the Board. On the one hand it professes to believe (with the Department of Energy) that the link between its own prices and world oil prices is weak; on the other, for unexplained reasons it evidently believes that overseas coal industries will take advantage of rising oil prices to increase their prices, thus establishing a strong link between imported coal prices and the world price of oil. On the face of it, the Board expects monopolistic behaviour by foreign coal producers but competitive behaviour from itself. To test the NCB's proposition, we comment briefly on the state of the world coal market.

Studies of international trade in coal indicate that, by the end of the century, there will be three large-scale exporters—Australia, the United States and South Africa—and a number of smaller-scale exporters such as Canada, Poland, the Soviet Union and China.[2] In the three major exporting countries

---

[1] The other large corporations in the British fuel market would also find their market power curtailed by relatively cheap coal imports.

[2] *Coal—Bridge to the Future, op. cit.*, p. 110, gives a list of coal export potentials in the year 2000 by country.

there is diversified private ownership of coal reserves, and production costs are relatively low. These conditions suggest that the international coal market will remain competitive in the long run.

The extent of competition in world coal is discussed in a study on steam coal prospects to the end of the century published in 1978 by the International Energy Agency. It points to the potential for expanding world trade in coal and suggests that coal prices will probably increase less than oil and gas prices

'. . . since supply curves are believed to be relatively flat in those OECD countries which have the most potential for expanded production (United States, Australia and Canada) and competition is thought to be sufficient to keep prices approximately in line with costs'.[1]

The nature of competitive forces in world coal and the scope for government action to promote more competition are explained in the Agency's study:

'In the absence of the limitations on the use of imported steam coal that currently exist in some countries, there is considerable competition in the world coal industry. Competitive forces in the international coal markets are enhanced by the widespread nature of the abundant reserve base, a lack of corporate concentration, few institutional constraints to entry (with the possible exception of access to capital), and competition from the suppliers of other forms of energy. . . . Government policies to promote freer trade and investment in steam coal could also enhance and ensure greater competition in world coal trade.'[2]

A powerful case for liberalising world coal trade is made by the Agency, which suggests that, with unchanged energy policies, both Europe and Japan will expand their coal imports substantially by the end of the century; the bulk of the increase is likely to be supplied by the United States, Australia and South Africa. For Europe, imports in the year 2000 are estimated at about 200 to 300 million tonnes, compared with 55 millions in 1976. European coal imports could, according to the study, rise to nearly 470 million tonnes in the year 2000 (compared with estimated European coal production in that year of 400 million tonnes) if a policy of removing coal trade barriers was pursued and the necessary transport infrastructure was built.

[1] *Steam Coal, op. cit.*, p. 16.   [2] *Ibid.*, pp. 55-56.

Like other studies of the prospects for world coal trade, the International Energy Agency's views on the speed at which it is likely to develop may well be over-optimistic. Very large investment in mines, inland transport, port facilities and ships will be required; there are 'environmental' problems to be resolved; and, not least, governments must be willing to remove import barriers. Nevertheless, there is little doubt that one of the most important responses of the energy market to the oil 'crisis' of the 1970s will be a growth in international trade in steam coal.

All the evidence we have seen suggests that the world coal market is reasonably competitive, though nowhere near to perfect competition. By contrast, the British coal market, in the absence of competition from imported coal or some other fuel, is comparatively monopolistic; British coal is also relatively high-cost (Section IV). We cannot therefore understand the NCB's argument that coal import prices will be 'greatly affected' by rising world oil prices whereas British coal prices will apparently not be so affected. Logic compels us to believe the opposite: competition in the world market will keep import prices down whereas British coal prices (which start from a relatively high level) are likely to increase sharply—unless constrained by more competition.

Government policy and coal imports

We would conclude that, if Britain is to benefit from a world-wide swing back to coal, the British fuel market must be open to imports. Protection of the indigenous coal industry from overseas competition would probably result in its uneconomic expansion since its monopoly power would be strengthened, and the NCB's costs and prices would tend to rise towards the ceiling set by oil prices.

. The NCB and the NUM are, for understandable reasons, opposed to coal imports. Protection against overseas competition would allow the British coal industry to capture much of the rent associated with rising fuel prices. For the community as a whole, however, the threat of imports would provide a valuable check on the Board's efficiency and enlarge consumer choice among sources of supply.

It might be supposed that the present Conservative Government would favour freedom for coal imports given the likelihood that, in the long run, world coal prices will rise more slowly than would British prices under a protectionist régime. At present there are no explicit restrictions on coal imports, although it appears that the CEGB and NCB have an understanding that imports of power generation coal will be limited.[1] But official publications seem generally unhappy about the idea of importing coal. At the Belvoir Inquiry, for example, the Department of Energy's *Assessment of Energy Requirements* used the concept of an 'import gap', carried over from the 1978 Green Paper and from *Energy Projections 1979*, in a rather contrived attempt to justify new coal investment projects. The Department's civil servants were placed in an awkward situation at the Inquiry by the NCB's failure to produce any demand estimates of its own for coal and other fuels (Section I). No doubt they felt some obligation to support the Board. Even so, the Department's case, which we examine below, seems surprisingly unsophisticated and lacking in economic content.

In essence, the Department's approach is to make projections of UK fuel consumption and set them alongside its estimates of indigenous fuel supplies. It then derives the residual difference, describing it as 'net fuel imports'. Since the residual is a relatively small difference between two much larger numbers, comparatively small proportionate changes in the demand and supply estimates will lead to much larger proportionate changes in the residual. Moreover, estimates of the demand for and supply of fuel late this century are subject to great uncertainty. Thus the potential for error in the residual import gap is enormous. To concentrate attention, as the Department of Energy does, on the difference between two numbers which are themselves highly suspect seems to us of little value as a guide to fuel policy. Indeed, this kind of 'gapology' may well be positively misleading.

As an illustration, we consider the Department's own calculations for the year 2000 in the 1978 Green Paper as compared with those in *Energy Projections 1979* (Table XVI). The 1979 projections revised downwards the *demand* estimates in the Green Paper, especially for the upper end of the range. The indigenous *supply* estimates were, however, reduced by a larger amount so that, instead of showing a 'gap' ranging from an export surplus of 65 million tonnes to an import requirement of 85 millions, there is a projected import gap of

---

[1] 'US exporters offer low-cost shipments of coal to CEGB', *Financial Times*, 30 September 1980.

TABLE XVI

ENERGY DEPARTMENT ESTIMATES OF UK
FUEL DEMAND AND SUPPLY IN 2000

	Green Paper 1978	Energy Projections 1979
	million tonnes coal equivalent	
Primary fuel demand	450 to 560	445 to 510
Indigenous supplies	475 to 515	390 to 410
Import gap	−65 to 85	35 to 120

35 to 120 millions. To the uninitiated, the Department appears to be making a prediction that there will be substantial net imports of fuel late this century. In reality, the quality of the supply and demand forecasts is such that the difference between them is barely meaningful. We have already commented on the likely over-estimation by the Department of fuel demand. At the Belvoir Inquiry we also suggested that its indigenous supply estimates are probably too low, especially for North Sea oil and gas. It would be simple enough, therefore, to set out supply and demand estimates of our own which demonstrated the likelihood of a fuel export *surplus* in the year 2000.

But such games with numbers are of little relevance. Our criticism is not so much of the Department's forecasts themselves—though we note the considerable change in the import gap between the Green Paper and *Energy Projections 1979*, which is not satisfactorily explained—but of its methodology. We contend that *no-one* is capable of making demand and supply forecasts for the year 2000 of sufficient precision for the gap between them to be meaningful.

Furthermore, even if net fuel imports could be forecast with reasonable accuracy, it is unclear why such a figure should occupy the central place in energy policy-making which it seems to be assigned by the Department of Energy. There is no reason why import minimisation *per se* should be pursued as a policy aim. The import gap concept concentrates on the sheer *volumes* of indigenous output and imports, whereas the relevant questions concern their comparative *costs*. It is much more important, in a consideration of the Coal Board's plans, to assess the likely competitive position of British coal than to concentrate unduly on whether the latest hypothetical calculations of supplies and demands happen to show an export surplus or an import gap in the year 2000.

We do not suggest that the Department of Energy would necessarily aim to promote the development of high-cost indigenous resources. However, the philosophy underlying its import 'gapology'—which seems to be that indigenous energy supplies should be available for almost any eventuality—may lead indirectly to import restraint. If investment is sanctioned in energy sources which turn out to be high-cost in relation to imports, there is a fair chance that when the sources come into production they will be subsidised. If, for example, the NCB opens pits in the Vale of Belvoir or elsewhere which are uncompetitive with imports, there will be strong pressure on government to keep the mines open—by persuading the CEGB to burn more coal, by providing more subsidies for the NCB, or by similar means—so that imports are in effect excluded. A reduction in fuel imports as a consequence of investment in indigenous resources which are low-cost relatively to imports— as is the case with North Sea oil—is clearly desirable. But we are very doubtful, for the reasons given in this *Paper*, whether the NCB's expansion programme will result in the production of coal which is competitive by international standards. The NCB has provided no evidence, either at the Belvoir Inquiry or, so far as we know, elsewhere, that the coal from its proposed new pits could compete with imports.

NCB's arguments against imports 'illogical'

The NCB's arguments against importing coal seem illogical. First, it suggests that coal import prices will increase as world oil prices rise but fails to acknowledge that British coal prices will be similarly affected. The NCB's second argument—that the balance of payments will suffer from coal imports—is incorrect in such circumstances. If imported coal is less expensive than British coal, it is wasteful of British resources and indirectly disadvantageous to the balance of payments to produce something at home which can be more cheaply obtained abroad. Similarly, the NCB's argument about employment advantages in developing indigenous coal relates to

[82]

employment in the coal mining industry only. It takes no account of the indirect effects on employment generally of misallocating resources. If a more efficient allocation of resources resulted from allowing coal imports to increase, employment opportunities in the economy as a whole should be larger. The last NCB argument—that there are 'strategic' advantages in supplying coal from indigenous sources—presumably means that British supplies are more secure than those from abroad. Many industrialists would disagree after the serious shortages produced by industrial action in the British coal industry in the winters of 1971-72 and 1973-74. In any case, security of supply is always a matter of degree and there is a limit to the premium consumers are prepared to pay for extra security. It is clearly in the NCB's interests to exclude imports, but its arguments that it would be disadvantageous for the nation seem to us quite without substance.

The Department of Energy also adopts an import-minimising stance, being concerned in particular with the country's 'security of supply' (page 81). A reasonable objective of national energy policy would be the lowest energy cost consistent with any selected level of security. There is, however, a trade-off between energy cost and security. Extra security can be bought, up to a point, but a government has to consider whether or not the likely benefits of reducing insecurity outweigh its costs. A sudden supply interruption can, of course, cause serious economic and social disruption, since in the short run the possibilities of substituting other factors for energy, or one fuel for another, are limited. Thus the costs of an interruption seem obvious.

Indigenous supplies are, however, also subject to interruption, and it is by no means clear that British coal is a more secure source of fuel than imports. British fuel policy, like the fuel policies of other countries, is apparently dominated by the belief that, in an uncertain world energy climate, it must be beneficial to promote the development of indigenous energy sources, almost irrespective of cost. But there are other security-increasing options that are already being pursued—such as holding strategic stocks of oil and the International Energy Agency agreement to implement temporary demand restraint and oil-sharing in times of emergency. These options seem likely to be more cost-effective than a policy of protecting British coal and do not appear to carry the same disadvantage of, in effect, forcing the economy on to a long-term growth path constrained by high-cost energy supplies. As we have argued in this *Paper*, coal industry protection is a very costly option because one of its side-effects would most likely be to increase the monopoly power of the industry and to channel into it rent from rising energy prices. Compared with such costs, the 'benefits' of a small quantity of extra indigenous coal production with uncertain security characteristics seem to us small and possibly non-existent.

It seems to us that growing international trade in coal will be one of the market's principal responses to rising oil prices, and that coal imports into Britain will, if the government allows, expand. Such an increase seems desirable. Actual or threatened imports would tend to hold down NCB costs and to limit rises in British energy prices. They might well also improve the security of our energy supplies by diversifying sources of coal. Whether NCB sales would be raised or reduced by unrestricted imports is uncertain. The direct impact would be to reduce sales, but the indirect efficiency-stimulating effect should be the opposite.

If Britain is not to participate in the growth of world coal trade, some very good reasons should be sought from the government. Both theoretical reasoning and the evidence strongly suggest that the alternative policy of maintaining the NCB's near-monopoly of the British coal market would impose costs on society which are clearly excessive in relation to any likely benefits.

APPENDIX

Future Demand for Coal by the Electricity Supply Industry

Colin Robinson and Mark Tomlinson

THE AUTHORS

CoLIN ROBINSON (see page 11).

MARK TOMLINSON was educated at Farnborough Grammar School and the University of Surrey, obtaining a First Class Honours degree in Physics with Business Economics in 1977. Postgraduate work, supported first by the Social Science Research Council and then by Shell Petroleum Company, concerns econometric modelling of residential energy demand patterns. In 1980 and 1981 he was appointed as consultant for industrial development models in Tunisia and Saudi Arabia and he is now Editor of *Energy Report*.

1. *Introduction*

In the autumn of 1979, we prepared for the Belvoir Inquiry estimates of coal consumption for power generation to the end of the century. This Appendix brings up-to-date those estimates, the principal conclusion of which was that the power generation market for coal would decline substantially by the year 2000, possibly to 45 to 65 million tonnes from 90 millions in 1980.

There are several reasons why our earlier estimates need to be modified. The most important concerns the likely rate of increase of electricity consumption. When we prepared our estimates for the Inquiry, the CEGB's forecast of the probable rate of increase of electricity consumption in England and Wales was about 2 per cent a year. To establish comparability with the CEGB's forecasts, we took 2 per cent a year as our 'surprise-free' projection of electricity sales in Great Britain and used four other scenarios to demonstrate the effects of altering key variables, such as the electricity growth rate and the size of the nuclear programme. In February 1980, towards the end of the Inquiry, the CEGB reduced its medium-term electricity growth forecast to 1 per cent a year (3 below). At the same time it produced for the Inquiry substantially reduced projections of its fuel consumption to the end of the century, based on the 1 per cent figure for electricity growth.

In this Appendix we have reduced our 'surprise-free' projection of electricity sales to 1 per cent a year in order to maintain comparability with the CEGB's forecast.[1] Thus we continue to assume that the CEGB's forecast is as reasonable an approximation as can be found at present to electricity market trends in the absence of major surprises. The electricity growth rates in our other scenarios (4 below) have also been reduced compared with those we used for the Belvoir Inquiry.

If lower electricity growth was the only change to our Belvoir Inquiry estimates, our calculations of future coal consumption by power stations would also be reduced significantly. Other changes, however, may compensate the coal industry, in whole or in part, for lower-than-expected electricity sales. For instance, recent relatively slow progress with the nuclear power programme suggests it will be smaller at the end of the century than we assumed at the

---

[1] Whether the 1 per cent electricity growth rate is consistent with our real GNP growth assumption of 2-2½ per cent a year (Section III) is uncertain. Between 1973 and 1980, however, electricity consumption increased at about half the rate at which real GNP grew.

time of the Inquiry, even though we thought then that the programme would fall well behind government expectations. This Appendix assumes a lower nuclear capacity at the end of the century than did our figures for the Inquiry.

Finally, oil consumption at power stations may be less in the 1980s than we thought because, as oil prices have risen in the last two years, the 'technical' minimum oil burn at the CEGB's power stations has turned out to be lower than it expected. This relative price effect should not, however, affect estimates of oil and coal consumption late this century—unless it is believed, along with the Department of Energy, that British coal prices will in the long run stay very depressed relatively to oil prices.

The purpose of this Appendix is to explore, as systematically as we can on specified assumptions, the outlook for coal sales to the electricity supply industry. It goes without saying that no-one can know with accuracy how this market will develop over the next 20 years. Our objective is simply to identify trends and quantify them as best we can so as to determine whether NCB and Department of Energy estimates are likely to be of the right order of magnitude.

2. *The Determinants of Power Station Consumption*

Over two-thirds of the coal consumed in the UK is sold to the electricity supply industry (Table II). About 90 per cent of what the industry takes goes to the CEGB and most of the rest to the South of Scotland Electricity Board (SSEB). The North of Scotland Hydro Electric Board (NSHEB) buys no significant quantities, and only about half a million tonnes a year are consumed in electricity generation in Northern Ireland. This Appendix is concerned with Great Britain only.

Provided they are left free to choose, the three electricity supply authorities in Great Britain will make decisions about their three main types of plant (coal, oil and nuclear) as follows:

Decisions about the construction of new power stations will be based on expectations about relative total costs (capital and operating) over the anticipated full lives—which may be up to 30 years—of the power stations.

Decisions about the operation of existing power stations will depend mainly on relative operating costs. That is, once a power station has been built, its capital costs are unavoidable and its utilisation rate will depend primarily on how cheap it is to run compared with other stations. The CEGB has a control system, the 'merit order', which ranks its power stations in order of operating costs; to minimise costs, the lowest-cost stations run on base load, those with somewhat higher costs have lower utilisation factors, and so on. The main element in operating costs is fuel.

[88]

Three factors appear to be critical in determining the amount of coal which power stations may require in the future:

(i) The expected rate of growth of electricity demand, which will determine the total capacity the electricity supply industry plans to construct.
(ii) The expected total costs of coal, oil and nuclear power stations, which will determine their respective shares of total capacity.
(iii) The relative prices of coal and oil.

It is reasonable to assume that, since nuclear stations have relatively low running costs, they are in the foreseeable future always likely to be placed higher in the merit order than coal or oil stations. Consequently, coal and oil stations, once constructed, can be regarded as being in direct competition so that their utilisation rates will vary inversely with relative prices. If, as in 1979-80, oil prices increase relatively to coal prices, the electricity supply industry will quickly promote coal-fired and relegate oil-fired stations in the merit order so that more coal and less oil will be consumed. If, as in the late 1960s and early 1970s, coal prices rise relatively to oil prices, coal consumption will fall and oil consumption will rise.

Figure A.1 shows how the utilisation rates of coal and oil stations have varied since 1968-69 and plots the relative prices of coal and oil. The inverse relationship between the coal/oil price ratio and the coal utilisation rate is clear if not particularly close in statistical terms, since factors other than fuel prices enter merit order calculations (such as relative thermal efficiencies of stations, transmission costs, and government pressure to burn coal) and the system can be significantly disturbed by accidents or strikes which reduce the availability of particular power stations or fuels. The utilisation rate of oil stations fluctuates more than that of coal stations, presumably because the former are a much smaller part of the system.

The analysis which follows begins by considering the likely rates of growth of electricity demand and generating capacity. It then discusses possible utilisation rates of coal capacity so as to examine what may happen to coal sales for power generation. For simplicity, only coal, oil and nuclear power stations ('steam plant') are considered; in 1978 steam plant provided 98 per cent of the electricity generated by the electricity supply industry, the other 2 per cent coming from hydro-electric plants, gas turbine and diesel stations.

3. *Electricity and Generating Capacity Growth to 1990*

Since the energy 'crisis' of 1973, electricity consumption in the UK has increased very little: between 1973 and 1980 the average compound rate of increase was only about 0·3 per cent a year. Forecast

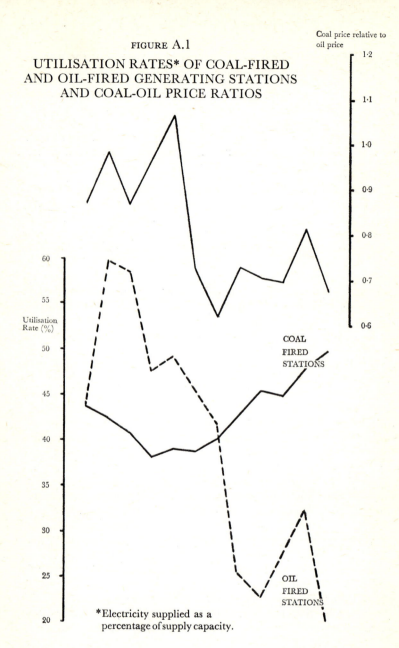

FIGURE A.1

UTILISATION RATES* OF COAL-FIRED AND OIL-FIRED GENERATING STATIONS AND COAL-OIL PRICE RATIOS

Coal price relative to oil price

1·2
1·1
1·0
0·9
0·8
0·7
0·6

60
55
Utilisation Rate (%)
50
45
40
35
30
25
20

COAL FIRED STATIONS

OIL FIRED STATIONS

*Electricity supplied as a percentage of supply capacity.

1968/9 69/70 70/71 71/72 72/73 73/74 74/75 75/76 76/77 77/78 78/79 79/80

Sources: Coal and oil prices:
Digest of UK Energy Statistics, Department of Energy.
Utilisation rates:
Digest of UK Energy Statistics, and *Annual Reports* of CEGB, SSEB and NSHEB.

[90]

TABLE A.1

OUTPUT CAPACITY OF STEAM PLANT*
ELECTRICITY SUPPLY INDUSTRY IN GREAT BRITAIN

| | | *Megawatts sent out*
(% of total capacity) | | |
	Nuclear	*Oil*	*Coal†*	*TOTAL*
1978-79	5,530	10,400	44,820	60,750
	(9·1)	(17·1)	(73·8)	(100)
1990	11,000	15,000	42,000	68,000
	(16·2)	(22·0)	(61·8)	(100)

*Excluding hydro, pumped storage, gas turbine and diesel stations.

†Including dual-fired and mixed-fired stations.

Sources: *Digest of UK Energy Statistics 1979* (Department of Energy); *CEGB Annual Report and Accounts*, 1978-79 and 1979-80; *CEGB Statistical Yearbook*, 1978-79 and 1979-80; *SSEB Report and Accounts*, 1978-79 and 1979-80; *NSHEB Report and Accounts*, 1978-79 and 1979-80; *CEGB Corporate Plan*, 1978; Electricity Council, *Medium Term Development Plan, 1979-86.*

electricity growth has also been adjusted downwards. Estimates in the *CEGB Annual Report and Accounts*, 1978-79 (para. 153) and the Electricity Council's *Medium Term Development Plan*, 1979-86 (para. 40) suggested rates of increase of about 2 per cent a year to 1985-86, both in maximum demand for electricity and in units sold; in the 1979-80 *CEGB Annual Report and Accounts* (para. 92) the forecast annual rate of increase to 1986-87 was cut to just under 1 per cent. Press reports[1] indicate there may be a further reduction in the forecast, and the CEGB is closing or 'moth-balling' over 3,000 MW of coal and oil plant early in anticipation of low electricity growth in the 1980s.

Any steam plant capacity which could be available to meet growth in demand to the mid-1980s is already under construction. Thus, it is possible to gain a reasonable idea from announced plans how the electricity supply industry's capacity is likely to change in the period to 1990. There is little prospect of bringing in new steam capacity before 1990, unless site work is begun in the very near future, since coal and oil plant have been taking about 10 years to complete and nuclear plant even longer.

In 1978-79 almost three-quarters of the output capacity of the steam plant was in coal-fired power stations (Table A.1). Table

[1] *Financial Times*: 'Demand falls as capacity rises', 10 September 1980; 'CEGB to "Mothball" 22 back-up power stations', 12 September 1980; 'Why 15,000 MW more is planned', 26 November 1980; and 'Electricity industry cuts its power demand forecast', 30 March 1981.

A.1 derives estimates of steam plant capacity in 1990 from plans already announced by the electricity supply industry (in the *CEGB Annual Reports and Accounts 1978-79 and 1979-80*, the *CEGB Corporate Plan 1978* and the *Electricity Council's Medium Term Development Plan 1979-86*). It assumes that the Advanced Gas Cooled Reactors (AGRs) now under construction at Dungeness, Hartlepool and Heysham will all be fully operational in the early 1980s and that the two AGRs on which construction work should begin in 1981 (the new Heysham station, and Torness in the SSEB area) will be operating at half capacity in 1990. The considerable increase in oil-fired capacity in the 1980s is a consequence of decisions taken some years ago; several large oil-fired power stations are being commissioned in the early 1980s. Only one new coal-fired station (Drax B) is under construction; allowing for delays, it will probably be in service by 1986 or 1987.

The large increase in generating capacity which would result from the existing construction programme is likely to be reduced by plant closures. The closure programme already announced by the CEGB (p. 91) will most likely be followed by more plant retirements in the 1980s; otherwise, substantial excess capacity would probably result. It is assumed in Table A.1 that around 7,000 MW of coal-fired and oil-fired plant will be closed in the 1980s, approximately what is required if the growth of capacity is to be no faster than the CEGB's growth of electricity sales.[1]

Allowing for both new construction plans already announced and plant closures, the total generating capacity of steam plant may be in the region of 68,000 MW by 1990 (Table A.1). The composition of total capacity seems likely to change significantly, with the share of coal dropping from about 74 per cent to a little above 60 per cent and the shares of both nuclear and oil increasing so that together they will have well over one-third compared with their present one-quarter.

4. *Growth of Demand and Capacity Beyond 1990*

Beyond 1990 the uncertainties are much larger since the electricity supply industry will have more flexibility than it has for the 1980s in deciding how to meet expected increases in demand. If site work were begun on new plant in the next two or three years, it should be in operation soon after 1990. It appears at present that both the electricity supply industry and the Government favour an expansion of the nuclear power programme. The CEGB's 1978 *Corporate Plan* (para. 5.7.1), for example, states:

[1] In our estimates for the Belvoir Inquiry we assumed about 3,000 MW of closures in Britain in the 1980s. The reduction in forecast growth of electricity sales and the closure plans already announced by the CEGB and SSEB make the earlier closure estimates too low.

'Nuclear power continues to be the most economic choice for electricity generation, taking into account both capital and revenue costs, and provides a robust long-term solution to the problems of new generating plant orders.'

Similarly, the *Electricity Council Medium Term Development Plan 1979-86* (para. x) states:

'The industry agrees with government energy policy that nuclear power must be developed as a bulwark against the decline of gas and oil supplies, especially since on current estimates nuclear stations appear to be the most economic choice for new generating plant.'

The Government subsequently endorsed the plans of the electricity supply industry in a statement by the Secretary of State for Energy (18 December 1979) in which he said:

'Looking ahead, the electricity supply industry has advised that even on cautious assumptions it would need to order at least one nuclear power station a year in the decade from 1982, or a programme of the order of 15,000 megawatts over 10 years. The precise level of future ordering will depend upon the development of electricity demand and the performance of the industry, but we consider this is a reasonable prospect against which the nuclear and power plant industries can plan.'[1]

In addition to the installation of more nuclear capacity, there is a possibility that one new large coal-fired station will be constructed (Grimsby has been mentioned), and others may be 're-furbished'. No new large oil-fired stations are at present planned. Opinions are, however, apt to change, given the uncertainties of the future, so the rest of this Appendix explores some of the possibilities by setting out a number of scenarios which attempt to present internally-consistent pictures of alternative futures for fuel consumption by power stations.

Scenario 1—No surprises

It is useful to begin with a 'surprise-free' scenario.[2] Although historical evidence suggests that surprises (compared with people's expectations) always occur, the 'no surprises' projection provides a valuable benchmark against which the impact of surprises can be measured.

In this scenario (Table A.2) it is assumed that growth in electricity demand proceeds fairly steadily at about 1 per cent annually from 1978-79 to the year 2000. We assume that, after 1990, the electricity

---

[1] *House of Commons Hansard*, 18 December 1979.

[2] For discussion of scenario construction and of the concept of 'surprise-free' projections, H. Du Moulin and J. Eyre, 'Energy scenarios: a learning process', *Energy Economics*, April 1979; and Robinson and Morgan, *North Sea Oil in the Future, op. cit.*, Chapter 3.

TABLE A.2

OUTPUT CAPACITY OF STEAM PLANT*
ELECTRICITY SUPPLY INDUSTRY IN GREAT BRITAIN:

'NO SURPRISES' SCENARIO

	Nuclear	*Megawatts sent out* *(% of total capacity)* *Oil*	*Coal†*	*TOTAL*
1990	11,000 (16·2)	15,000 (22·0)	42,000 (61·8)	68,000 (100)
2000	22,000 (29·3)	13,000 (17·3)	40,000 (53·4)	75,000 (100)

*
† } Notes as for Table A.1.

supply industry builds sufficient plant to keep up with the growth of demand so that plant capacity reaches 75,000 megawatts sent out (MWso) in the year 2000. To provide the extra capacity of some 7,000 MWso between 1990 and 2000, the industry builds the 15,000 MW of nuclear plant mentioned in the Government's statement of December 1979 (p. 93). However, since it is assumed that 4,000 MW of ageing 'Magnox' capacity is closed in the 1990s, nuclear capacity in the year 2000 is only 22,000 MW compared with the 'maximum availability' figure of 40,000—which now seems far out of reach—mentioned in the 1978 Green Paper on Energy Policy (Cmnd. 7101, para. 10.18 and Annex 1, para. 5). The net increase in nuclear capacity exceeds the net increase in total capacity in 'no surprises' because we assume that, when plans for capacity in the 1990s are made, the electricity supply industry still believes, as it does now, that nuclear power will be the cheapest method of generating high-load-factor electricity. To avoid excess capacity, we assume that 2,000 MW of coal plant and 2,000 MW of oil plant are closed in the 1990s. Including Magnox, total plant closures are therefore 8,000 MW.

As a result of the specified assumptions, coal capacity falls by 2,000 MW from 1990 to 2000 and its share of total capacity falls from just over 60 to about 53 per cent.

Scenario 2—Expanded nuclear programme

This scenario (Table A.3) assumes that the electricity supply industry, either of its own volition or prompted by government, decides in planning for the 1990s that there is a serious risk of sharper rises

[94]

TABLE A.3

OUTPUT CAPACITY OF STEAM PLANT*
ELECTRICITY SUPPLY INDUSTRY IN GREAT BRITAIN:

'EXPANDED NUCLEAR PROGRAMME' SCENARIO

Megawatts sent out
(% *of total capacity*)

	Nuclear	*Oil*	*Coal*†	*TOTAL*
1990	11,000	15,000	42,000	68,000
	(16·2)	(22·0)	(61·8)	(100)
2000	25,000	11,500	38,500	75,000
	(33·3)	(15·3)	(51·4)	(100)

*
†} Notes as for Table A.1.

in the prices of fossil fuels than it is currently forecasting. Consequently, it opts for a bigger nuclear programme and closes more coal and oil plant. Total capacity in the year 2000 is the same as in 'no surprises' but nuclear capacity is 25,000 instead of 22,000 MW. Despite the delays to the nuclear programme, 25,000 MW appears to us to be just attainable by the end of the century. Assumed retirements of coal and oil capacity increase from 2,000 MW each in 'no surprises' to 3,500 MW each in this case. In consequence, oil capacity falls to about 15 per cent and coal capacity to about 51 per cent of the total. Nuclear's share rises to one-third.

Scenario 3—Nuclear delays

In this scenario it is assumed that, because of further public opposition and/or technical problems, it is not possible to achieve by the year 2000 the 'no surprises' nuclear capacity of 22,000 MW. Although the electricity supply industry *plans* for 22,000 MW in 2000, it achieves a capacity of only 15,000 MW. If there were to be delays to the nuclear programme, it is unlikely it would be possible to order sufficient coal- or oil-fired plant in time to compensate fully for the loss of nuclear capacity. In any event, the industry might not want fully to make up for the nuclear shortfall because of the likely higher relative cost of fossil fuel capacity which would raise electricity prices and restrict the growth of demand.

It is assumed here that oil capacity is 2,000 MW higher than in 'no surprises' and that coal capacity is 3,000 MW higher. Such increases could be achieved either by fewer closures or by new power station construction. Since electricity demand is assumed to

TABLE A.4

OUTPUT CAPACITY OF STEAM PLANT*
ELECTRICITY SUPPLY INDUSTRY IN GREAT BRITAIN:

'NUCLEAR DELAYS' SCENARIO

	Nuclear	Megawatts sent out (% of total capacity) Oil	Coal†	TOTAL
1990	11,000 (16·2)	15,000 (22·0)	42,000 (61·8)	68,000 (100)
2000	15,000 (20·5)	15,000 (20·5)	43,000 (59·0)	73,000 (100)

*
† } Notes as for Table A.1.

rise more slowly in this case (at about 0·7 per cent a year in the 1990s, instead of 1 per cent), the expansion of coal and oil capacity is sufficient to meet total demand. Total capacity is 73,000 MW compared with 75,000 MW in 'no surprises'.

Table A.4 shows that, in this case, the shares of nuclear and oil in total capacity might both be a little over 20 per cent, with coal's percentage share nearly 60 per cent (significantly higher than in 'no surprises').

Scenario 4—Fast electricity growth

This scenario assumes that, in the next few years, changes in the British energy market—but particularly rises in the relative price of natural gas—improve the competitive position of electricity. Electricity sales consequently start to grow faster than of late and the supply industry begins to plan for a rate of increase of about 2 per cent a year to the early 1990s (1 per cent a year under 'no surprises'). In the event there are capacity shortages by the late 1980s, since the installation rate of new plant cannot keep pace with the faster growth of demand; electricity prices begin to rise faster than in the 'no surprises' case; and the growth of demand for electricity declines a little. Even so, over the whole period from 1978-79 to 2000, growth of demand is about 1½ per cent a year and capacity eventually matches it, rising to 84,000 MW.

To achieve the extra 9,000 MW, as compared with 'no surprises', the electricity supply industry is assumed to plan a bigger increase in nuclear capacity since it expects nuclear power to be cheaper than coal or oil. We assume, however, that when plans are being made for the 1990s, the industry sees a significant risk that bottle-

OUTPUT CAPACITY OF STEAM PLANT*
ELECTRICITY SUPPLY INDUSTRY IN GREAT BRITAIN:

'FAST ELECTRICITY GROWTH' SCENARIO

	Nuclear	*Megawatts sent out* *(% of total capacity)* Oil	Coal†	TOTAL
1990	11,000 (16·2)	15,000 (22·0)	42,000 (61·8)	68,000 (100)
2000	25,000 (29·8)	15,000 (17·9)	44,000 (52·3)	84,000 (100)

*}
†} Notes as for Table A.1.

necks will hold nuclear capacity to a maximum of 25,000 MW in the year 2000 (as in 'expanded nuclear programme'). Thus, as an interim measure, the industry stops closing coal and oil plant and builds an additional 2,000 MW of coal-fired power capacity. The outcome (Table A.5) is that coal capacity is 4,000 MW higher than in 'no surprises' but its share of total capacity is slightly lower.

Scenario 5—Slow electricity growth

The final scenario considered is one in which, because of slow growth in the British economy and better energy conservation induced primarily by rising energy prices, electricity demand increases at an annual rate of only ½ per cent on average to the year 2000. If the nuclear programme was the same as in 'no surprises', large amounts of fossil-fired plant would have to be retired because total capacity in 1990 would be sufficient to meet demand in 2000. It is assumed (Table A.6) that there is an 11,000 MW reduction in fossil-fuel capacity in the 1990s (6,000 MW coal and 5,000 MW oil). Coal capacity falls to about 36,000 MW in this scenario and its share of total capacity is about 53 per cent (Table A.6).

The scenarios summarised

The results of the five scenarios in comparison with capacity figures for 1978-79 are illustrated in Figure A.2.

Coal capacities in the year 2000 under the five scenarios and their shares of the total are summarised below:

	Coal capacity (MWso)	Coal's share of total capacity (%)
1. No surprises	40,000	53·4
2. Expanded nuclear programme	38,500	51·4
3. Nuclear delays	43,000	59·0
4. Fast electricity growth	44,000	52·3
5. Slow electricity growth	36,000	52·9

In 1978-79 (Table A.1) coal capacity was about 44,800 MW and its share of the total nearly 74 per cent.

It would, of course, be possible to define a wider range of coal capacity in the year 2000 by taking more extreme scenarios. However, considerable care must be taken not to introduce internal inconsistencies into the scenarios. For instance, it would be illegitimate to specify a 'nuclear delays' scenario combined with 'fast electricity growth' since delayed installation of nuclear plant would probably result in higher fossil fuel prices than in the 'fast electricity growth' case, with the effect, other things being equal, of reducing the rate of growth of demand for electricity.

There are two important common features of the five scenarios. The first is the substantial reduction in the share of coal in total steam generating capacity as a consequence of growing nuclear capacity and, to a lesser extent, increasing oil capacity. The second is the rising share of nuclear power in the total. Because of the comparatively low running costs of nuclear stations, they are likely always to appear at the top of the merit order and to operate at high load factor. Consequently, as nuclear's share of capacity rises, the utilisation rates of fossil fuel power stations are likely to decline.

TABLE A.6

OUTPUT CAPACITY OF STEAM PLANT*
ELECTRICITY SUPPLY INDUSTRY IN GREAT BRITAIN:

'SLOW ELECTRICITY GROWTH' SCENARIO

	Nuclear	Megawatts sent out (% of total capacity) Oil	Coal†	TOTAL
1990	11,000 (16·2)	15,000 (22·0)	42,000 (61·8)	68,000 (100)
2000	22,000 (32·4)	10,000 (14·7)	36,000 (52·9)	68,000 (100)

*
†} Notes as for Table A.1.

[98]

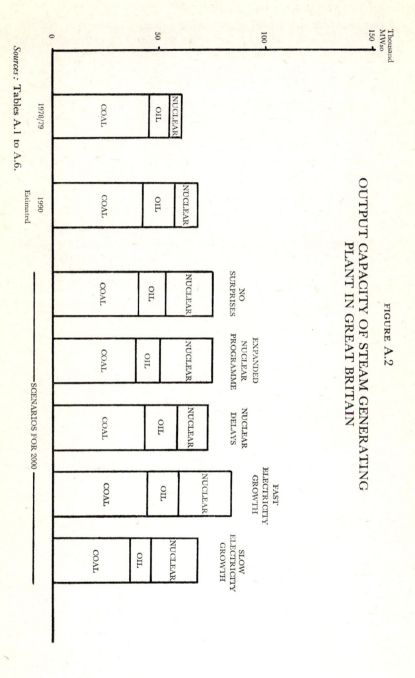

FIGURE A.2

OUTPUT CAPACITY OF STEAM GENERATING
PLANT IN GREAT BRITAIN

Sources : Tables A.1 to A.6.

[99]

5. Coal Generating Capacity and Coal Sales

The estimates of coal-fired generating capacity in the year 2000 in each of the five scenarios are now translated, by the following procedure, into estimates of quantities of coal sold for power generation:

Total electricity generated from fossil fuels

First, various utilisation factors (varying from 65 to 75 per cent[1]) are assumed for nuclear generating capacity in order to calculate a plausible range for the amount of electricity produced by nuclear stations. Electricity generated from fossil fuels is then derived as a residual by subtracting nuclear electricity from estimated total electricity supplied.

Electricity generated from coal

This process gives a range for total electricity supplied from coal and oil stations. To divide the total between coal and oil, a model derived from econometric analysis of experience in the 1960s and 1970s is used. There is a relationship between the relative prices of coal and oil and the utilisation rates of coal-fired and oil-fired power stations (Figure A.1). Indeed, we know that, because of the way the CEGB merit order system is operated, a change in one direction in, say, the relative price of coal (with other things equal) is quickly followed by a change in the opposite direction in the amount of coal used in power stations. Consequently, our model assumes that, with given coal and oil generating capacities, relative utilisation rates will depend on relative prices. If the price of coal rises relatively to that of oil, the utilisation rate of coal power stations will fall, and *vice-versa* for coal price reductions.

Our analysis of the past suggests that, for coal to be competitive with oil in power generation, its price must be significantly below that of oil. For the share of coal in the total electricity generated by fossil fuels to be equal to its share of fossil fuel generating capacity, it appears that the price of coal must be about 90 per cent of that of oil (in pence per therm) to offset the greater convenience in use of liquid fuels. This conclusion is consistent with the statement in the 1978 Green Paper (para. 6.13) that coal needs a price advantage of about 1 pence per therm (on a 1978 price of about 10p per therm) in existing power stations before it is preferred to oil. To calculate the effects of variations in the coal/oil price ratio, we have estimated from past statistics how the share of coal-generated electricity in the total electricity generated by fossil fuels tends to vary as the relative prices of coal and oil change. Six price ratios (coal price:

---

[1] 75 per cent is assumed in the Electricity's Council's *Medium Term Development Plan, 1979-86*, Table 9.

ESTIMATED FUTURE COAL USE FOR POWER GENERATION IN GREAT BRITAIN IN THE YEAR 2000

million tonnes

Relative price (coal ÷ oil)	0·7	0·8	0·9	1·0	1·1	1·2
Scenario 1	52-58	50-56	49-55	47-53	44-50	43-48
Scenario 2	46-54	45-52	44-51	42-49	40-46	39-45
Scenario 3	62-66	60-64	59-63	56-60	53-57	52-55
Scenario 4	56-62	54-61	53-59	51-56	48-54	46-52
Scenario 5	43-49	42-48	41-47	39-45	37-43	36-41

oil price) were used—0·7, 0·8, 0·9, 1·0, 1·1 and 1·2—to demonstrate the sensitivity of coal-fired generation to price. It seems to us unlikely that the price ratio will lie outside our chosen range except in circumstances where Britain has access to relatively cheap imported coal (Section VI).

Coal use for power generation

Finally, the estimates of coal-generated electricity in different scenarios and at different price ratios were transformed into estimates of coal use for power generation by assuming the following average thermal efficiencies of coal stations—33 per cent in 1985, 34 per cent in 1990 and 35 per cent in 2000. The results are given in Table A.7.

The range of estimates is wide, particularly for 2000—inevitably so since it reflects the uncertainty of the future. Nevertheless, it can readily be seen that NCB estimates of the future demand for coal for power generation (Table XIV) look extremely optimistic. The *lower* end of the NCB range—75 million tonnes in 2000—is well above the highest figures in Table A.7. The Department of Energy's estimates (66 to 78 million tonnes in 2000) also appear to be on the high side, although it should be noted that the Department assumed in *Energy Projections 1979* growth rates of future electricity demand of 1·7 to 2·3 per cent a year, which at the time were close to the CEGB's projections.

It might reasonably be concluded from Table A.7 that a significant decline in power station consumption of coal is likely by 2000; nearly all the figures fall within the range 40 to 60 million tonnes, or 30 to 50 million tonnes less than in 1980. Only if the electricity supply industry could import substantial quantities of low-priced coal would we expect coal consumption by power stations to be above the top end of our range.

[101]

QUESTIONS FOR DISCUSSION

1. 'The commitment we must have is that the country will sell every ounce of coal we produce' (Mr Joe Gormley, President of the National Union of Mineworkers). Evaluate this demand from the point of view of the economist.

2. Why are the prospects for the *British* coal industry gloomy in a world where rapidly rising oil prices have afforded coal a major competitive opportunity?

3. 'The British coal industry is a monopoly.' Discuss.

4. Evaluate the measures British governments have employed in the past to protect coal from competition from imports and other fuels.

5. 'The coal industry needs to be placed in a more competitive environment so that it can be more readily judged whether investing in new pits is a wise use of productive factors or a misallocation of resources.' Assess this proposition.

6. Why is there a strong probability that, with the coal industry protected from overseas competition, British coal prices would be linked to and rise in step with world *oil* prices?

7. Freed from the threat of coal imports, 'the British coal industry would capture most of the rent arising from the increase in prices of other fuels'. Explain.

8. Why is it bad economics to argue, as do the NCB and NUM, that government must support the coal industry with subsidies and import barriers to preserve jobs and help the balance of payments?

9. Which is likely to assure the greater 'security of supply' to British coal consumers: free access to foreign suppliers, or dependence on a domestic monopoly?

10. In the three major coal exporting countries—Australia, the USA and South Africa—there is diversified private ownership of coal reserves. Do you think this partly explains why they are low-cost producers? If so, discuss the reasons.

FURTHER READING

Coal's recent history and coal policy are discussed in three IEA Papers by Colin Robinson:

A Policy for Fuel?, Occasional Paper 31, 1969.

Competition for Fuel, Supplement to OP 31, 1971.

The Energy 'Crisis' and British Coal, Hobart Paper 59, 1974.

The National Coal Board's plans are described in:

Plan for Coal, 1974.

Coal for the Future, 1977.

The Department of Energy's views on coal are in:

Fuel Policy, Cmnd. 2798, HMSO, 1965.

Fuel Policy, Cmnd. 3438, HMSO, 1967.

Energy Policy Review, 1977.

Energy Policy: A Consultative Document, Cmnd. 7101, HMSO, 1978.

Energy Projections 1979, 1979.

Various studies of the world coal market have appeared in recent years, for instance:

OECD, *Steam Coal: Prospects to 2000*, 1978.

World Coal Study: *Coal—Bridge to the Future*, Ballinger, 1980.
Future Coal Prospects: Country and Regional Assessments, Ballinger, 1980.

International Energy Agency, *Report of the IEA Coal Industry Advisory Board*, December 1980.

Those who require statistics of the coal industry should consult the NCB's *Annual Reports* and the very useful Department of Energy publications:

Digest of UK Energy Statistics (annual);

Energy Trends (monthly).

First published in May 1981

© THE INSTITUTE OF ECONOMIC AFFAIRS 1981

All rights reserved

ISSN 0073-2818
ISBN 0-255 36143-2 ✓

Printed in England by

GORON PRO-PRINT CO LTD
6 Marlborough Road, Churchill Industrial Estate, Lancing, W. Sussex

Text set in 'Monotype' Baskerville

What Future for British Coal?

*Optimism or realism on
the prospects to the year 2000*

COLIN ROBINSON

and

EILEEN MARSHALL

University of Surrey

Published by

THE INSTITUTE OF ECONOMIC AFFAIRS

1981